Praise for *Goats Giving Birth*

Anyone new to goats should read this book! Deborah Niemann explains the process of goat birthing with clarity through many examples of what can go right or wrong. She teaches us to, above all, keep a level head.

—Rebecca Sanderson, lifetime country girl, writer,
Goat Journal, *Backyard Chickens*, and *Countryside & Small Stock Journal*

This book is as engrossing as it is invaluable to anyone anticipating goat births. Deborah Niemann's practical guide describes real-life experiences to complement any textbook knowledge—because so many births are not textbook cases. Instructive and sometimes poignant stories are accompanied by helpful photographs and enlightened advice.

—Tamsin Cooper, goatwriter.com

A delightfully excellent and informative kidding resource complete with great photos for new and experienced goat owners alike!

—Katherine Drovdahl, MH CA CR CEIT DipHIr QTP,
and author, *The Accessible Pet, Equine and Livestock Herbal*

It's the most wonderful (and frightening) time of the year! From normal births to problem deliveries, all new goat owners could use an experienced mentor during kidding season. *Goats Giving Birth* is like having a long-time goat-owning friend waiting at your fingertips.

—Marissa Ames, editor, *Goat Journal* magazine

If you are starting out keeping goats and forming a breeding program, you need this book. It's full of wisdom, practical experience, and reality. Be ready with advice from a long-time goat keeper. There is no sugar coating between the covers, although there are plenty of great photos to add to the written word. Niemann tells it like it is, the good, the bad, and the heartbreaking, but with good reason. As a goat owner interested in breeding, you need to know this important information. Grab a copy and settle in for a well-crafted read that travels through the years of a respected goat keeper's life.

—Janet Garman, TimberCreekFarmer.com,
author, *50 Do-It-Yourself Projects for Keeping Goats* and
The Good Life Guide to Keeping Sheep and Other Fiber Animals

Birth: both thrilling and terrifying to novice and experienced herdsmen. So much can go wrong—and right! Through actual accounts of goat births from normal to difficult, *Goats Giving Birth* prepares readers for the many ways that goats bring life into the world. Highly recommended.

—Karen Kopf, Kopf Canyon Ranch

Goats Giving Birth is such a useful book for anyone who plans to raise goats and wants to be prepared for labor and delivery. But it's also useful for those of us who have been tending to goat births for many years! The pictures are high quality and extremely educational. The vast experience held in these pages will prepare you for just about any possibility you might encounter as your doe's midwife. Even though I've helped bring well over 100 goat kids into the world and have nurtured my share of weak babies, I learned so many new tips and sage pieces of advice from this book.

—Kate Johnson, author, *Tiny Goat, Big Cheese,* and writer for *Goat Journal*

Goats Giving Birth

Goats
giving birth

what to expect during
kidding season

Deborah Niemann

new society
PUBLISHERS

Cover design by Diane McIntosh.
Cover images ©iStock.
Photo credits: Katerine Boehle pp. 4–13, 47–56.
Printed in Canada. First printing June, 2020.

Inquiries regarding requests to reprint all or part of *Goats Giving Birth* should be addressed to New Society Publishers at the address below. To order directly from the publishers, please call toll-free (North America) 1-800-567-6772, or order online at www.newsociety.com

Any other inquiries can be directed by mail to

New Society Publishers
P.O. Box 189, Gabriola Island, BC V0R 1X0, Canada
(250) 247-9737

LIBRARY AND ARCHIVES CANADA CATALOGUING IN PUBLICATION

Title: Goats giving birth : what to expect during kidding season / Deborah Niemann.

Names: Niemann, Deborah, author.

Description: Includes index.

Identifiers: Canadiana (print) 20200211870 | Canadiana (ebook) 20200211889 | ISBN 9780865719422 (softcover) | ISBN 9781550927344 (PDF) | ISBN 9781771423311 (EPUB)

Subjects: LCSH: Goats—Parturition—Popular works. | LCSH: Goats—Parturition—Anecdotes. | LCSH: Goat farming—Anecdotes. | LCSH: Kids (Goats)—Anecdotes.

Classification: LCC SF383.N52 2020 | DDC 636.089/82—dc23

Funded by the Government of Canada	Financé par le gouvernement du Canada

New Society Publishers' mission is to publish books that contribute in fundamental ways to building an ecologically sustainable and just society, and to do so with the least possible impact on the environment, in a manner that models this vision.

Contents

Introduction

There is nothing about goat ownership that creates more anticipation, excitement, frustration, and fear than birthing. It's wonderful to walk into the barn one morning and see a couple of kids bouncing around and nursing. But it feels like you've been punched in the stomach when you walk in and see a distressed doe or a dead or malformed kid. If you've never seen a goat give birth before, you don't really know if something is normal or not. As a former childbirth educator and doula, I knew all about human birth, but I soon learned that goats are very different.

Only three months after I brought home my first goats in 2002, I became a member of several goat groups on Yahoo. Because I knew no goat owners, other than the woman who sold me the goats, the Yahoo groups filled the role that a knowledgeable neighbor or parent would have filled a century ago. Whenever something happened that worried me, I'd sign on and ask for help. There were always other goat owners out there in cyberspace who offered advice and encouragement.

Today there are also Google groups, Facebook groups, and a host of other groups. In 2009, I started my own group on Ning for owners of Nigerian dwarf goats (nigeriandwarfgoats.ning.com). Through the years, I have seen thousands of posts from goat owners all over the world, and I've noticed that kidding is the event that causes more anxiety than anything else.

So many people join an online group and post something like this:

We're new goat owners and awaiting the birth of our first kids! Anything we need to do or know? How do we know everything is going okay? What do we do if we have to help? Any advice is appreciated! Thanks!

They may also sign on and post something like this:

Our first goat has been in labor for two days, and we're worried! What should we do?

This book is part of the answer to those questions, but because every birth is different, it is also useful for those who are not new to kidding. In my book *Raising Goats Naturally: The Complete Guide to Milk, Meat, and More*, I explain all of the technical aspects of birthing goats. But most goat owners don't feel that information is enough preparation because there are so many exceptions to "normal" and so many of them are not problematic. Knowing textbook averages is not very helpful. The average length of time for the first stage of labor for a first freshener is twelve hours, but three hours is normal, and eighteen hours can also be normal. Just as some women are in labor for two hours and others for two days, goats can be different from one another.

When someone asks me if something is normal, my answer is almost always "It depends," and then I ask a dozen questions or more as I try to understand that particular situation. Every goat is unique, and every birth is different, even from one year to the next with the same doe. I have some goats that were born on this farm, gave birth for ten years and are now retired. Each birth was different even though they could usually be described as "normal."

In this book, I share stories of my goats giving birth. I want to make my experience your knowledge—not because I've experienced everything but because the more you know, the more you realize you don't know. I describe the experiences bluntly and share my thoughts and fears and confusion at the time. The pictures, too, illustrate the reality of these births. My hope is that these stories will take the place of being at the birth and add to your knowledge base and confidence in yourselves but more importantly in your goats. When you're new, the big overarching fear is that your goats can't do it. You have to help—or even worse, you believe you have to save them. This is almost never true. According to *Goat Medicine* by Mary Smith and David Sherman, 95 percent of births require no assistance, and the longer I have goats the more I believe this number is accurate.

The idea behind this book is to share stories of real goats giving birth to help you sort out which 5 percent actually need your assistance. I view every kidding as a learning experience, from the very first one to the ones that will happen in our next kidding season. In addition, I learn from experiences that other people share with me, and I sometimes think of those stories when I am with a goat in labor. Many of the stories in this book come from my Antiquity Oaks blog, where I chronicled the births of our goats from 2006 to 2016. Most of the posts were written within a day or two of the kidding, so you get to read all of the raw emotions—the joy and sometimes the sadness—that came with each birth.

The advantage of reading the birth stories here is that in addition to the original story of the birth, I give you my assessment of the birth as I understand it today with accumulated knowledge and experience. I look back on some of these births and know that I should have done something differently. With other births, I realize that nothing would have made a difference in the outcome. As you read through these births, you will probably be thinking about what you would have done in that situation and how you might have responded differently. At some point, you might also get paranoid—or you might think that there is something terribly wrong with our goats and that none of this will ever happen to you.

The book starts with a variety of "normal" birth stories so that you can see the wide range of what's normal. Please do not be tempted to skip over these "boring" stories. Often when someone has never seen a goat give birth, they think something is wrong when it isn't. Sometimes the doe is not even in labor yet! I knew one woman who spent several nights in the barn and finally called her vet out, only to learn that her two does were not even pregnant. In fact, the most common mistake that I see is someone assuming something is wrong when everything is perfectly normal. Birth takes time, and unfortunately, that gives us plenty of time to worry, which means plenty of time to do something wrong.

Stories of difficult births are presented in chapters "Not So Normal Births," "Caesarean Section Births" and "Death." I purposely share the stories of our worst births. We have had more than 650 kids born on our farm as I write this, and the vast majority of births have been happy occasions with smiling humans and healthy kids. We have had only two caesarean sections, and only two does have died as a result of kidding complications. There are farms with worse records, and there are farms with better records, but if you have goats long enough, you will have some unhappy experiences. However, the happy experiences will far outweigh the sad ones. Whenever anyone asks me what I love most about our farm life, I always respond, "Kidding season!"

Back L–R Nina, Scarlet.
Front L–R Sadie, Giselle.
All pregnant!

Disclaimer

Keep in mind that I am not a veterinarian and that nothing in this book is meant to be veterinary advice. It can be challenging to figure out what to do when you are in the midst of a birthing situation, and it would be a mistake to assume that any information provided in a book represents the path you need to take in your particular situation. This book is intended to be an educational tool to help you better understand the normal and not-so-normal experiences that you may encounter when goats give birth. It's like all of those childbirth books we humans read when we're pregnant.

If you are ever worried about what's happening during a goat's labor or birth, you should call an experienced goat veterinarian. Please do not call your dog's or cat's vet and ask them to see your laboring goat. As one small animal vet said to me many years ago, "Believe me when I tell you that I would not be doing you any favors by seeing your goats." Experienced livestock vets are usually happy to discuss a birthing situation with you over the phone and even talk you through some procedures or help you decide when it's time for their professional assistance.

Lizzie licks one of her newborns while it learns to nurse. This is the happy outcome in 95 percent of births.

NORMAL BIRTHS

Although 95 percent of goat births don't require intervention, you don't know if your 5 percent will happen after you've had fifty or eighty goats give birth or if it will happen when your first goat gives birth. I have one friend who had two or three does for fifteen years before she had to assist in a birth. On the other hand, someone bought two does from me, and the first birth ended with both kids dying when they were about nine hours old because they never nursed and she didn't know that was a problem. The second birth involved a kid that needed assistance to be born and the doe died a few hours later.

But if you have never seen a normal birth, how do you know if the goat in front of you is acting normally? It is fairly common for new goat owners to think that something is wrong when everything is going fine. It seems that if you could just see a birth or two, you wouldn't worry as much. Right? Chapters "Normal Births" and "Normal But Different Births" tell stories of normal births and how "normal" humans respond. You will see that it is normal for us to wonder if everything is okay, even after having seen quite a few goats give birth. After a few years, my motto became "If the goat is happy, I'm happy," and even after eighteen years of seeing goats give birth, I still chant that in my head when I start to wonder if everything is okay.

Cleo's twin doelings

B·L·O·G
WED
APR 14
2010

We had a visit last weekend from Sarah, our apprentice from November. She came back because she wanted to see a goat give birth. Starting on Thursday, Cleo's ligaments were so soft that I kept thinking she was going to give birth "pretty soon." When I left for a speaking engagement in Chicago on Thursday, I figured she'd give birth later that night. When I left to pick up pigs on Friday, I figured she'd give birth while I was gone. When I got home with the piglets, however, Mike and Sarah said that Cleo had been waiting for me.

I went into the barn and sat down on the straw with her. She gave me more kisses than I've ever had from any goat. She licked my face and my neck over and over as I sat with her in the kidding pen. She kept making little two-syllable "ma-a, ma-a" bleats. She kept looking at my lap and pawing at my legs. I could tell she was thinking about crawling into my lap. She would lie down next to me on one side, and then almost immediately, she would get up again, turn around, and lie down on her other side. She was clearly uncomfortable. I went to the walnut grove where Mike and Sarah were finishing repairs on the fence before releasing the piglets into their new home.

"Cleo is getting close," I said to Sarah. "You don't need to hurry, but I'm not sure if I'll be able to come get you later."

Sarah came with me, and when we got back to the barn, I could tell that Cleo was very serious about giving birth. She was no longer making the little bleating sounds. Instead, a whispery moan escaped her throat with each push. She lay on her side and pushed her legs out in front of her body. Her big belly almost caused her to roll onto her back, but she jerked and pulled herself upright again.

"No matter how many times you see this, you always get to a point where you feel like it's taking too long," I said to Sarah. "But really, she's fine. There's no sign that anything is wrong."

Finally, a hoof started to peek out, then a second hoof. "This is exactly the way it's supposed to be," I said. "First the front hooves, then the nose." And as if it were scripted, a nose appeared. "This is a textbook birth." The whole head appeared, and the body quickly followed. I put the little doe up by Cleo's face, so she could help me clean it off.

While Cleo licked her baby, I wiped it with a towel. The little doe shook her head and sneezed. Within minutes, she was scooting around the straw performing the goat baby equivalent of crawling. Cleo stood up and lay down a couple times. Then she seemed to stare off into the distance as if concentrating on something that none of us could see. I said to the little doe, "Okay, kid, you're on your own. It's time for mommy to birth another baby." And the second kid was born quickly.

Two does! Of the sixteen kids born so far this year, twelve are does. When you raise dairy goats, that's the equivalent of winning the lottery. Of course, we are only halfway through kidding for the year, and things could turn around, but I'm enjoying the dozen little does in the barn at the moment. And yeah, I'm keeping one of these.

Usually Cleo was a very aloof doe. She was not a cuddly goat the rest of the year. But I always knew when she was in labor because she suddenly became the friendliest doe on the farm. This birth occurred after we'd been raising goats for eight years, and, thankfully, I was learning patience by then.

It is important that there not be a big audience when a doe is in labor. Over the years, we have had a lot of interns during kidding season, but it has never been more than one at a time, and the goats usually get a chance to know the intern before giving birth. It's important for goats to feel safe when they are in labor, or their contractions may not be productive. Remember, they are prey animals and are always wondering if a new stranger is going to eat them. Twice we've had does go into labor during an open farm day, and in both cases, labor was unusually long and the doe didn't give birth until almost everyone was gone. In the case of the second doe, I was ready to take her to the university vet clinic and told my husband

Sherri cranes her neck as she begins to push out the second kid while the first kid lies next to her.

Three separate bubbles begin to emerge, and you can see white hooves in the top one.

to get a dog crate loaded into my car as soon as the event ended. Luckily, it took him forty-five minutes, because when he came back to tell me it was ready, she was pushing, which saved me a two-hour drive—and the experience of delivering baby goats in my car! So, as tempting as it is to invite guests for your goat births, it's not a great idea.

✦ ✦ ✦

The next two births are both with Sherri, the fifth doe that I purchased when starting my herd. She was retired after kidding when she was 10 years old and then enjoyed 6½ more years in the pasture with her daughters before passing away peacefully while napping one afternoon. She never had any kidding problems. The irony, however, is that having seen fewer than a dozen goats give birth by 2005 when she first kidded on our farm, we thought she was having difficulties, which I describe in the first post below. I often think of Sherri's first birth on our farm when new goat owners are worried about a goat that they think is in labor.

It's triplets!

B·L·O·G THU FEB 16 2006 — Sorry I haven't posted in a few days, but I am recovering from four days in bed with the flu. Last night, however, I got a welcome back to the real world gift of three new baby goats! When I went to the bathroom at 1:00 a.m., I heard a goat over the baby monitor. There is only one reason a goat makes noise in the middle of the night, so I grabbed my clothes and got dressed as quickly as I could. Sherri was at day 149 as of midnight, and normal gestation is 145–150 days for Nigerian dwarfs. I grabbed a big stack of clean towels and ran out into the unseasonable thunderstorm and headed for the barn. I wasn't more than twenty steps into the barn when I heard the familiar squeak of a newborn kid. I ran up to Sherri's stall to see one baby on the ground while mama was working on bringing the second one into the world. I dropped to my knees to start drying the one already born, keeping my eyes on Sherri the whole time. When the big bubble of fluid popped and a head emerged, I placed the first baby under the heat lamp and caught the second baby as it was sliding into the world. In the middle of drying off that one, I realized I didn't know if they were boys or girls, so I looked between their hind legs, thrilled to discover that both were does!

Sherri stood up so casually I wondered if she would be having only two this year, even though she's always had triplets before. As she licked her two little daughters, the gold and white one started bopping her mama's chest, stomach, neck, etc., looking for her first meal. After a few minutes, Sherri plopped down again, and I'm not even sure that she made a sound as she easily gave birth to baby number three,

The head is out, but the sac is still intact. If it doesn't break or get broken before the umbilical cord breaks, the kid could suffocate.

I use my fingernails to rip the amniotic sac and pull it off the kid's face as he is being born. I would have done it a little sooner, but Sherri was pushing him out too fast!

I'm continuing to pull membranes off the face and wipe off mucus so that the kid's airway is clear and he can breathe.

I place the second kid near Sherri's face so she can lick him and start to bond.

a buckling. Within a couple of minutes, she was once again standing and licking her babies. I stayed with them for about an hour until I started to feel weak and dizzy and a little sick to my stomach. I realized that I probably should not be spending too much time in the barn while I'm still recovering from the flu. I went inside and woke up my youngest to go sit with the babies to make sure everyone was nursing. Although the little gold doeling figured out the nursing thing very quickly, the other two were still pretty clueless.

Sherri's birthing this year was so much better than last year, even though 1:00 a.m. births are not my favorite. Her labor went exactly as last year's, but our reaction was different. Sherri spent two or three days really looking like she was in labor. She'd lie down and push her legs out in front of her, or she'd squat and push. It really looked like she was in labor. Last year, I posted a message on a Yahoo goat group, asking for advice after two days of Sherri's unusual labor, and most people responded with all sorts of dire possibilities. My daughters were quite worried and convinced we needed to intervene. Finally, we did a vaginal exam to discover that she had not even started to dilate. The next day, however, she gave birth to her babies in the pasture during the fifteen minutes when no one was out there!

This year, as I was lying in bed sick, the girls gave me reports from the barn regularly. Finally, it occurred to me that Sherri gives birth just like I do. It took me a day or two of labor with each of my three children. My body would putz around having contractions that irritated me and even hurt, but they didn't do anything to actually get the baby out. But when my body did finally decide it was time to give birth, they each came flying out in record time: I pushed for twenty-five minutes with my first, twenty minutes with my second, and one really big push birthed my third child. That's what Sherri is like. She putzes around for days, but when her body is ready, the babies come flying into the world. This year, I was determined to honor her unique way of giving birth.

Sherri stands up and continues licking the first two kids while a couple of bubbles hang out of her back end.

In addition to being a great example of a normal goat birth, this story illustrates how goats will give birth whenever they need to give birth. You may be asleep or in the middle of lunch (as the next story shows), and there will be days when it is not convenient for you. You may even be sick, but the goat is going to give birth if it's time. Many of us joke that goats purposely give birth at the most inconvenient time. Of course, that's not true. It sure seems like it, though!

Looking back on this now I really wish that I had a video of what I thought was "pushing,"

Sherri is still standing but kid number three gets closer to making his appearance. You can see him just beginning to present.

The head is out, and once again Sherri is pushing so fast that I'm not able to break the sac as quickly as I'd like to.

The kid's body is out, and I still haven't been able to break the amniotic sac.

because it clearly was not real pushing. Goats do not push for two days before giving birth to perfectly healthy kids. A goat is not really pushing until it is craning its neck, curling its tail over the back, and either holding its breath or letting out some kind of noise. I often think of Sherri's first birth on our farm when a new goat person tells me that their goat has been pushing for hours or that it has been in labor for days. I always ask them to explain exactly what they mean by that. Pushing is exhausting, and most does can't do it for more than a couple of hours. Then they just lie down and give up, and contractions get farther and farther apart until they stop entirely, which you will see in some of the later stories.

B·L·O·G THU MAY 13 2010 Sherri's triplets (for the sixth time)

Sherri is 7 years old, and she came to live here as a yearling. Her breeder said not to worry about kidding difficulties. As a yearling, Sherri kidded in the pasture with triplets while her owner was preparing a kidding pen for her. Well, "Don't worry" is subjective. It kind of depends on what you want to worry about. I do not have to worry that she'll have any sort of dystocia. Her pelvis is big enough for a Mack truck to go through at top speed. However, there are other things to worry about.

As a 2-year-old here, we thought she was in labor for two days, and we kept her in the barn. Then I finally decided that we had no clue what was happening, so we let her go out into the pasture. About fifteen minutes later, my son reported that there were three kids in the pasture with her, and one wasn't looking good. When I got there, I thought it was dead, but my daughter insisted it was alive. We took the tiny doeling into the house and put her in a sink of warm water because she was ice cold and clearly suffering from hypothermia. She finally snapped back into the world, and she grew up to be a fine doeling. I, however, am still traumatized by the experience five years later. Sherri gives birth too efficiently—so easily that she doesn't have time to clean off the three or four kids that she always births.

Every year, Sherri makes me think she's going to kid any minute now for about two days. This year (as in years past), I thought that I was older, smarter, wiser, and I'd know. Right? Wrong! For two days, I kept thinking that she was going to kid soon. On Saturday, although she didn't act like she was in labor, her belly was hollowed out between her ribs and hips and her ligaments were so soft that they could be gone any minute. Her udder didn't look like it was ready to explode, but there's a little wiggle room in that particular criterion. On Sunday, we had tickets to a Broadway play in Chicago, and I tried to explain the situation to Sherri, but she

I finally am able to dig my fingernails into the sac as another bag of amniotic fluid is coming out with the kid.

The toughness of the amniotic sac can vary from birth to birth. Most of them are not difficult to break, although it's easier to do if you are not wearing gloves. Being able to use your fingernails definitely makes it easier.

looked at me like I was nuts. "Could you please have your kids now? Within the next hour?" Nope. Katherine stayed home on kid watch. Even though Sherri wasn't showing signs of labor, she can go from zero to three kids in about twenty seconds. So Katherine waited... and watched... and waited... and watched... and you get the idea. Sherri did not kid Sunday while we were at the play.

Monday morning, the ligaments were gone, so I knew it would be soon. But the thing about Sherri is that she is the most stoic goat in the world. She makes NO sound until the kid is actually being born — as in, the kid is shooting out at that moment.

Logic and science not being on my side, I decided to trigger Murphy's Law to get Sherri to go into labor. I told Katherine that I was going to make some bran muffins. I figured that I'd be in the

I use my fingers like a squeegee to clean off the kid's nose so he can breathe.

middle of making muffins when Sherri would start to give birth, because that would be really inconvenient. I peeled and diced an apple, mixed up the muffins, put the muffin batter in the oven, and still no sound from the kidding barn. Fine! I'll have lunch! I heated up some leftover tamale pie that we'd had for dinner the night before, and I sat down. I took a bite…and "Maaaaaaaa!" came over the baby monitor. Katherine looked at me and laughed as I said with a mouthful of food, "ONE bite!"

"I'll check on her." Katherine said. A moment later, as I was in the middle of my second bite, I heard her scream over the monitor, "Baby!"

With my mouth full of tamale pie, I pulled off my reading glasses and dropped them on the table, dashed to the front door,

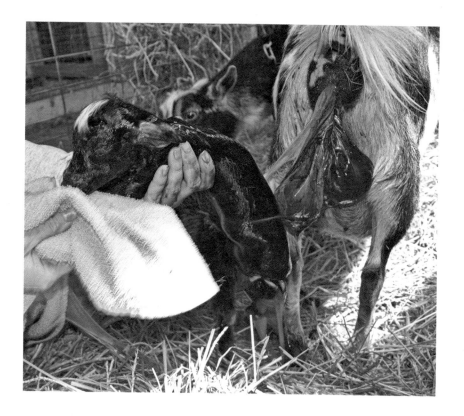

Then I use a towel to get even more mucus off his nose. Yes, Sherri gave birth standing up. It's not terribly common, but sometimes goats do it. When I saw the thick umbilical cord to the left of the membranes, I started to think that she might be done and there would be no more kids.

pulled on my shoes, and ran across the yard towards the barn. (Note: Chewing and running are really not compatible activities. Do NOT try this at home!) I arrived at the kidding barn as I was swallowing my tamale pie, trying not to inhale anything and choke.

Katherine was laughing about Sherri's impeccable timing, and I suggested that she retrieve her brand new camera from the house so that we could get pictures. (She's been saving her money for months to buy a fancy DSLR camera.) She took my advice and then proceeded to take more than one hundred photographs of the birth.

Sherri took an unusually long time between kids this year. For her, that means we were able to get each one dried off before the

One of Sherri's kids finds his first meal within minutes of being born! Unfortunately, the camera snapped before he really had a good latch on the teat. In this photo, he has only about half of it in his mouth, which would be uncomfortable for Sherri if he continued nursing like that. Luckily, most kids realize quickly that they need to have the entire teat in the mouth while nursing. The tongue should cover the lower teeth. You'll notice that he is not standing up but rather is on his knees to be able to access the teat. Only the tiniest of kids with the tallest of moms would stand up to nurse.

next one was born. It was really fun compared to most years, when she is shooting them out faster than we can dry them or even check the gender of each kid.

The first two kids were does, and then she had a buck, which was pretty cool because that is exactly what was reserved from her. So, yes, that means that all these sweet little darlings are going to another farm to live. But that's okay, because this was a repeat breeding of the one that produced Jo, whom I dearly love. She fed triplets last year as a first freshener, so she's an awesome little milker.

Back at the ranch, uh, house, the timer on the oven was going off. Luckily, Jonathan (my son) was still inside, and he knew it was for the muffins, so he pulled them out of the oven. No one, however, knew that I had put my big tomato plants out on the deck for their hardening off time. When I went back inside an hour later and saw them out there, they were only a little wilted, and I pulled them inside immediately.

Sherri is one of those goats who gave birth here nine times. Her second stage of labor was always faster than average. If we had not understood that every birth is different, there might have been times we would have been worried about her. In fact, when she was 10 years old, she was in labor when I got a call that some of our llamas were at a farm about half a mile away. We had an intern at the time, and I told her that I was going to go get the llamas because I wasn't worried about Sherri. The intern had already attended several births, so she knew the basics of drying off kids when they were born, and that's all that likely would be needed with Sherri. I told her that if by chance the kid wound up being in some weird position, it wouldn't be an emergency situation and that she could just call me and I'd come back. After about half an hour I got a call on my cell. The intern was worried that something was wrong because Sherri was pushing but the kid wasn't coming out. I told her not to worry, reiterating the goat's excellent birthing history and that we had plenty of time. It was obvious that she was very worried in spite of my attempt to reassure her, so I told her I was on my way back and that she should not do anything to intervene in the meantime. When I arrived, I found the intern toweling off a newborn kid. We named the doeling Sophie Kinsella because all of Sherri's kids were named after authors. Sophie still lives with us today and has the same excellent birthing ability as her mom.

Seeing the birthing ease that Sherri and all of her daughters, granddaughters, and great-granddaughters display is one of the reasons I am so committed to not interfering in every birth. I want to have goats that have the natural ability to give birth easily. I don't feel like I am doing the breed any favors by routinely doing a vaginal check on every doe, rearranging kids that are still high in the uterus, and then pulling kids. How do you know if a goat gives birth with ease if you never let her try to do it on her own? If you've read this far, you know that I want to be at every birth just in case, but if a doe repeatedly needs help birthing her kids, I won't continue to breed her because I don't want to perpetuate those genetics.

NORMAL BUT
DIFFERENT BIRTHS

The previous stories were textbook births. The following stories are not textbook, but they are not abnormal in a bad way. It is not usual for a first freshener to have quads, nor it is common for a doe to have three breech kids, but neither of these scenarios is necessarily bad, either. A lot of goat owners worry about breech births, but most of them go just fine, which we learned in our very first kidding season. The third doe to give birth on our farm was in the midst of birthing a breech kid, and my 10-year-old daughter and I panicked. While we were debating which one of us should go wash our hands to assist (because the book said you had to assist in a breech birth), the doe easily pushed out the kid. We learned quickly that we needed to adopt a practice of watchful waiting before deciding to assist in a birth.

B·L·O·G
THU
APR 15
2010

Bonnie's big baby bonanza!

When I checked Bonnie's tail ligaments yesterday, they were soft, so I kept checking on her every hour or sooner if I happened to be near the kidding barn. We had just finished rooing* a yearling Shetland ram, and I was about to start evening chores when I heard a sound that caught my attention. Was it Bonnie? When I looked into the window, I realized the loudmouth

---

* Some Shetland sheep have a genetic trait that produces a coat that can be hand plucked instead of being sheared. The process of hand plucking the coat is called rooing. This is done late winter/early spring. We've had only a couple of sheep that could be rooed.

was Andi (nothing new there), but Bonnie looked like she was in labor, even though she was quiet. I walked into the barn and positioned myself so I could see her back end. There was a big blob of bloody mucus under her tail. Yep, definitely in labor. I walked into the kidding pen and sat down across from her. She stood up, walked to the far corner of the pen, and stared at me.

"You don't want me here?" I asked. "That's okay, I'll leave."

I checked everyone's water buckets and gave the goats more hay, keeping an eye on Bonnie the whole time. When she was lying down and pushing beyond the point of no return, I stepped into the pen and sat down. Once a head emerged, I crawled toward her, ready to wipe off the kid's nose. As soon as the baby was born, I laid it next to her face so she could start to clean it off. But instead of starting to lick it, she curled her lip up and continued pushing. What? I had only cleaned off the nose and didn't even know the sex yet, and another head was emerging.

I grabbed another towel, popped the bag covering the second baby's head and started to wipe off the nose, worried that a third

When you get a kid started on the bottle, you have to open their mouth, put the nipple in there, close the mouth, and then hold it while the kid figures out that this is how they get milk. The first few feedings may take ten or fifteen minutes just to get an ounce into a kid because it's just dripping out of the nipple as the kid swallows it. If you get lucky, a kid may start sucking right away once they get their first taste of milk, but most take a day or two to get good at taking a bottle. If a kid has nursed from a doe, it can take much longer for them to get accustomed to taking milk from a bottle. Never assume that a kid is not hungry if it refuses a bottle.

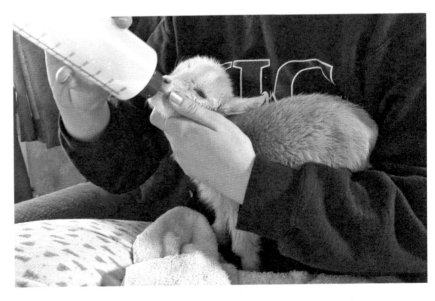

kid would be shooting out even faster. Moments later, kid number three was born, and I laid it on the towel next to Bonnie's face. She started to lick the two kids in front of her, and I finally was able to check the sex. The first kid was a buck, and the second one was a doe. Since Bonnie was quite busy with the first two, I decided to clean up the third one myself. It was a buck.

Less than five minutes later, Bonnie pushed again, and I saw another bubble emerging. Triplets are unusual for a first freshener. Usually they have only one or two, but Bonnie had been larger than normal, so I had already been thinking about the possibility of triplets. I'd be nuts to think she had a fourth one in there, although I wasn't completely sure what was hanging out of her back end. Was it the cord that's attached to the placenta, or was it membranes attached to another baby? After ten minutes, not having seen another kid, I convinced myself it was just the end of the placenta.

When Bonnie stood up, I thought, *Yep, she's done.* And then— plop! A little bag of mucus fell out of her back end onto the straw. I wasn't even sure if it was a kid, but I started clawing at the membranes with my fingernails to pop the bag. I could feel bones, so I knew it was a kid, but it was so tiny, and there was no color. Everything was beige. I ripped at the membranes, pulling them off everything until I found a nose. I wiped it off but couldn't feel any movement. Maybe it's been dead for a week or two, and that's why it's so small? I kept wiping. Then a sneeze and a shake of its head told me the kid was alive.

I put a clean towel down on the straw and laid the tiny kid on it. *Maybe this is going to be one of those tiny little kids that just doesn't make it,* I thought. *Don't get your hopes up. Don't get attached. It is just so darn cute and small and helpless. Of course, it's a doe. Seems like they're always the tiny*

When I let the doelings have the run of the house, they would curl up in Joy's dog bed and fall asleep next to my desk because they thought I was their mom.

We bought a playpen when a daycare was going out of business and started using that for kids in the house.

ones. It would never have a chance to get enough food with three big siblings and a mama that only has two teats. And the mama is a year-ling. Maybe in a couple of years, I'd trust Bonnie to raise four kids, but not this time. I picked up the other doeling and laid her on the towel with the tiny one. I placed both bucklings next to Bonnie's face so she could continue cleaning them and bonding with them.

It's not fair, but in the world of dairy, girls are worth more. Since Bonnie is a first freshener, the little bucks will be wethered and sold as pets. They're simply not worth as much, so I'll be raising the does, while Bonnie learns to be a mommy with the boys. And in the real world, little goats have to be able to hustle to get lunch, and I'm less convinced of the tiny doe's ability to do that. The first time we were confronted with a situation like this, my husband asked if we were messing with nature's plan by coddling a smaller, weaker kid. What about survival of the fittest? At the time, I had no answer other than simple compassion. I couldn't let a kid die. After eight years of seeing little goats grow up to be big and strong, I don't think that birth status is necessarily a big indication of ge-netic potential. Sometimes kids just get unlucky in utero and wind up with the short end of the placenta. Carmen and Coco, my first two house goats, are perfect examples of runts growing up to be big, strong, and healthy. But, I digress…

Back in the kidding barn — the bigger doeling was already trying to stand, while the little one was still learning to hold up her head. Membranes were hanging out of Bonnie's back end exactly as before the fourth one had been born, but I was fairly confident that there were no more kids. Bonnie seemed very content to clean the two babies in front of her, so I wrapped up the girls in a clean towel, hugged them to my chest, and headed for the house.

"Bonnie had quads!" I yelled to Jonathan as I walked in the door. As we set up the playpen for the doelings, I thought about how lucky it was that last weekend we saved some of Cleo's colostrum. The day after her kids were born, her udder looked full, so we milked her and put the colostrum in a freezer bag marked with her name and the date.

While the colostrum was thawing, I decided to milk Bonnie to see if I could get a little more, but her boys had already sucked down one side to nothing, and I barely got enough from the other side to cover the bottom of the bucket. The little boys are very tenacious, though, so I'm sure they'll get plenty. I took Bonnie's colostrum back into the house. It was an ounce. Cleo's colostrum had thawed enough that I was able to add two more ounces to the bottle, which would be plenty for their first meal.

Bottle-feeding is so complicated compared to nursing. Kids just know how to nurse. It might take them a few minutes, but they get it figured out quickly. Bottles, however, can be tricky. Colostrum poured all over my hand and the little doeling's face. The hole in the nipple was too big. I tried another nipple but had the same

If you have a fabric playpen, you need to put something waterproof on the bottom. Otherwise the whole thing will absorb urine and stink. Then your only option is to take it outside and hose it down on a sunny day. (Do I have to tell you how I know this?) We usually put a plastic trash bag on the bottom, covered by several layers of newspaper, then a towel on top.

problem! Pritchard teats don't have a hole in them when you buy them, so I pulled out a brand new nipple and barely cut off the tip to make the tiniest hole imaginable. It worked, but the little does still didn't know quite how to do it. It's twenty hours later now, and they still don't quite have the hang of it, but I'm sure that they'll be sucking like pros before the day is done.

Joy the bichon and Porter the English shepherd are both convinced that the babies are theirs. Joy stares and whines and paces next to the playpen. She barked at a guest who arrived this morning—she has never barked at anyone, but I suspect she wants to protect her babies. When I give the babies their bottles, Porter is right there to clean the dribbles off their faces, and he growls at Joy if she gets too close, even when they're in the playpen, so I can't leave the two of them unsupervised around the kids. One reason I don't like bottle-feeding is that I get too attached to the babies, even if they are the brattiest goats in the barnyard. And this little one is just so cute and cuddly. But I really don't need to keep more does this year! I really don't need to!

During our first eight years, we learned the hard way that does cannot always feed all of the kids that they give birth to. Sometimes it's because they can't produce enough milk. Sometimes it's because one of the kids isn't aggressive enough to get its fair share of milk from its hungry siblings. It seems like a terrible joke of nature that an animal with only two teats can give birth to more than two kids while a cow, which normally has one baby, has four teats.

First fresheners produce less than older does, which is why I was not even going to give Bonnie a chance to feed all four kids. Had she been an older doe with a good milk production record, I might have taken only one of the kids to bottle-feed, but some first fresheners cannot even produce enough to feed triplets. Combine that with the fact that it's better for a bottle baby to have a goat friend than to be an only child in the house, and it was an easy decision to bottle-feed two of the kids.

Today we weigh all kids at birth and then daily for the first couple of weeks. We want to see an average weight gain of 4 ounces per day for each Nigerian dwarf kid. That means that if it gained only 3 ounces one day, we want to see 5 ounces the next day. If it continues to gain less than 4 ounces per day, we know it's not getting enough milk. We rarely have this problem today, however, because we've learned how many kids a doe can feed, depending upon her age, her genetics, and her milking history. It is much easier to get a kid started on a bottle at birth rather than a few days down the road.

B·L·O·G TUE FEB 22 2011 Bonnie's quads

If I was a bettin' woman, I would have lost big last week.

Bonnie was the least wide of the four goats that were due to kid. She had birthed quads last year as a first freshener, but she had been much bigger—at least, that's what I remembered. So, I was pretty sure she was only going to have twins. As each day passed and she didn't give birth and she looked wider, I started to think maybe triplets.

Mike had made fajitas with homemade tortillas on Valentine's Day, and I had *just* taken the last bite of my first one when we heard a goat over the baby monitor. I gave Mike a look. I suppose I should be thankful that I had eaten one fajita and didn't have to run out there with nothing more than the aromas of delicious food lingering in my nostrils.

At least it was warm enough that I didn't have to deal with the heating pad and blow-dryer. With temperatures around freezing, a heat lamp and plenty of towels would be enough. When I walked into the barn, I saw something already poking out under Bonnie's tail. I ran into the office and grabbed the whole stack of towels without bothering to count. When Mike walked in a few minutes later, I was drying off the first kid. A few minutes after that, a second kid popped out. "You know, she doesn't really look any smaller," I said to Mike. "Maybe she's having triplets."

And then this tiny little thing shot out! Bonnie didn't seem to notice. "It's another kid!" I broke the amniotic sac and pulled it off the kid as I said, "I think it's dead," because it wasn't moving at all. I laid it on a towel and started cleaning it up. It shook its head and sneezed. "Oh, it's alive! But it's so tiny!" My excitement over the kid was short-lived, though, because Bonnie let out a bleat, and I looked down to see another kid presenting. Mike was at Bonnie's head and couldn't see what I saw.

"Here, take this one! Another one is coming! Gimme a clean towel!" I handed him the tiny kid in the towel as I grabbed a dry towel from him just in time to catch kid number four. "Quads again! I can't believe it!"

We sat there for a while making sure everyone was dry and nursing. And everyone was up and bouncing around in no time except for the littlest doeling. Seeing her next to her siblings, I was curious about the weight difference, so I weighed her and the biggest buckling. She was 1½ pounds, and her big brother was just over 4 pounds. I had a feeling that she wouldn't have a chance against her siblings when it came to getting fed. In the ultimate proof that life is not fair, goats have only two teats, yet they can have so many babies! Although she did manage to stand a few times, walking was clearly not her forte. In the meantime, the biggest buckling was already getting more than his fair share of mama's colostrum. It seemed he was always on one teat, while the other doe and buck took turns on the other one.

Luckily, Giselle had just kidded with twins twenty-four hours earlier. Normally, when a goat has a single or twins, we milk them at twenty-four hours and put it in the freezer for emergencies or other situations where we need more colostrum than what the mama goat is producing. If I could have turned back time, I would have milked Bonnie right away, before the big buckling had made such a piggy of himself. Although it wouldn't be a problem to have the three kids nursing practically non-stop, because they'd get a little each time, milking her for the tiny one would be useless because we wouldn't get enough to put in a bottle. So, the decision was made to bottle-feed the baby with Giselle's milk.

What a week it has been with the tiny doeling that Katherine is calling her Little Valentine. We brought her into the house so that we could give her the frequent, small feedings that she needed. She didn't usually take more than an ounce at a time for several days. On the second day she drank 2 ounces for a couple of feedings, and we were getting excited, but then she reverted to 1 ounce a feeding on the third day, and her little mustard poops had streaks of blood. I had never seen that before, so asked some other goat breeders. One of them, who bottle-feeds all of her kids, said that she had seen it a few times and it probably was not a big deal. She called the vet, freaking out the first time, and the vet listed several reasons why it could happen, and said that it was not a problem. Because most of my kids are raised by mama, I don't see every single poop, which would explain why I'd never seen it before.

The whole week with her has been one step backward after every two steps forward. She took 3 ounces at one feeding, but then reverted back to 2 ounces. I finally reminded myself that "normal" is a subjective term. This is her normal. It really does not matter that most kids at a week are sucking down 4 ounces in a couple of minutes. She is just a little delayed, and there is not anything that anyone can do about it. Patience has never been one of my stronger virtues, but I'm learning.

By the time Little Valentine was a month old, it was obvious that she was a normal kid who simply had had a rough start in utero. She grew normally and caught up with her siblings in size. Bonnie went on to have quintuplets a couple of years later!

In more than one story you will note that I didn't know if a kid was alive but I was cleaning off its nose and rubbing it dry with the towel anyway. I am always on auto-pilot with every kid that's born, assuming it's alive. It takes so little effort to clear a kid's airway and start to dry it with a towel while waiting for it to exhibit signs of life. Most of the time a kid will squeal or try to lift its head the moment it's born, but if a kid is even slightly compromised, it may not have that much energy. If you've cleared the nose and have been briskly rubbing it for half a minute or more, you can stop for a few seconds to see if you can find a pulse by placing your fingers on the kid's chest. You can easily feel the heart beating if it is alive.

◆ ◆ ◆

Although most kids are born nose and front hooves first, a fairly high percentage of kids are born breech, which is tail first in the veterinary world. Hind legs first is technically called posterior and is not considered an abnormal presentation. As long as the kid is still entirely in the mother's body, there is usually nothing to worry about with either of these presentations. However, once the umbilical cord is out, it may be compressed or broken, which means the kid's supply of oxygen has been cut off. The kid's head should then be born quickly so it can start to breathe. The following birth is unusual because all three kids were born breech. However, everything went fine, even though we were also dealing with sub-zero temperatures. Combine the freezing weather with Sherri's ability to give birth so quickly, and you can understand why this post started with me talking about being paranoid. This was also our first kidding season in our newly repurposed kidding barn. It has a heated office with windows that look into the kidding pens, so we can watch does in late pregnancy and in labor while keeping ourselves warm up until the last minute.

Three breech bucks

B·L·O·G
THU
JAN 15
2009

My diligence finally paid off, and I'm glad I was so paranoid. Shortly before 11:00, Mike and I were out in the office and I decided I wanted to clean up the pump room, which is connected. After about fifteen minutes, I thought I should look in on Sherri, even though she hadn't made a sound. When I looked out the window, I could only see her back end, but I knew immediately that she was pushing. I told Mike to get the girls. I put two towels under my coat so they'd stay warm, put on latex gloves under my mittens, and went to Sherri.

Shortly after the girls arrived, Sherri screamed. Margaret laughed and said, "That's what you usually hear from your bedroom in the middle of the night, right?" I said, "Yep, that's it, so we should see a kid soon." But there was no kid. Instead there was another contraction and another scream. After another contraction, I said, "Gee, I didn't think it took me that long to get dressed and get out here, normally." Sherri is so stoic, she normally doesn't make a noise until the first kid's head is being born, and by the time I open the barn door, that kid is already screaming. After a couple more contractions, Katherine said she saw a hoof.

Sherri is 6 years old and has never had any trouble kidding, so as long as the kid was lined up straight, she should be able to get it out. I was hoping we'd see a nose soon, since sometimes the nose is just little bit behind the first hoof. Nope,

The colder it is when you use a blow-dryer, the closer the dryer must be to the kid's body. If it's too far away, the air will be cold by the time it hits the hair and it will do no good. I am always rubbing the kid's hair with my hand so that I know whether the air is hot enough—or too hot—and I can adjust the distance accordingly. Plus fluffing up the hair gets the air blowing down to the roots for complete drying all the way to the skin.

no nose. Margaret crawled around to Sherri's back end and said, "That's a tail." Katherine argued, "No, it's a hoof." And they went back and forth. Finally, I crawled through the straw to Sherri's back end and said, "Well, unless there is hair growing on the hoof, it's a tail, because I see hair." As Sherri pushed, it became obvious that it was indeed a tail. The kid emerged slowly, inch by inch, as we commented on how big it was. Finally, it was born. As soon as the hips emerged, the rest of the body quickly followed. "It's a buck," Katherine said, as I was trying to dry it. Then I handed him to Margaret, who was sitting near Sherri's head, so she could blow-dry it while Sherri licked it.

I realized that the latex gloves really provided no protection for my hands, as they were rapidly going numb, so I pulled the gloves off and said I needed to get a dry pair. It was only slightly better than having my hands wet. When I ran into the office to grab more gloves, Katherine yelled that the second kid was presenting. As I headed back out, she said, "It's a hoof." When I got there, I realized that the hoof was upside down. I was worried that it was front hooves, which would mean that the kid was upside down. Katherine thought it was the back hooves. I'm glad she was right, because it was much easier for Sherri than the alternative. And it was another buck.

We were barely able to start drying off the second kid when the third one was sticking his little butt into the world. "Another breech?" I screeched. But that one slipped out with little effort on Sherri's part—at least it didn't appear to take much effort.

When a kid is first born, I have the heating pad in my lap covered by a towel to absorb the wetness. If it's cold enough to be using a heating pad, I'm usually freezing and enjoy the added benefit of getting my legs warmed up a little. Plus this keeps the bottom of the kid warm while I'm blow-drying the top. When a kid is completely dry, I lay them on the heating pad where they stay until they're ready to get up. Once they are standing, the heating pad is no longer doing any good, so I remove it from the stall. I don't leave the heating pad in there unless I'm there because I don't want to risk a goat chewing on the cord and getting electrocuted or starting a fire.

Of course, Sherri just *had* to give birth in the corner opposite where the heat lamp was hanging. I tried to get her to walk over to the heat lamp before the first kid was born, but she wouldn't budge. After the kids were born, we knew we had to get them over there because the blow-dryer wasn't working fast enough for three kids to get dry and stay warm. The girls carried the three kids to the lamp, hoping Sherri would follow, but she didn't. I picked up a kid and took it back to her and let her lick it, then took a step away so she'd have to take a step to lick it again. Eventually we got her to the heat lamp.

We thought we had the kids dry, but a few minutes later, Margaret said an ear of one of the kids was frozen. I checked the other two kids and found another one with a frozen ear. "That's frostbite! It's when your extremities freeze!" Everything I knew (or had forgotten) about frostbite was rushing through my head—*warm up quickly? Don't warm up too quickly?* Crap! Not knowing whether I was right or wrong, I decided to hold the kid's ear between my fingers and rub it; then I started to think that rubbing was wrong, so I just held it. Within a couple minutes it started to feel flexible again.

The kids seemed to have no idea they should nurse, so I decided to give them some Nutri-Drench, which is this stuff that is mostly molasses. When Sherri saw it, she started licking the syringe and then started sucking on it. She had been shivering badly and didn't seem to be helped by the two towels we had draped over her, so I thought the sugar might help her and I let her suck down several syringes of the sticky brown liquid. I wished I had my 60 cc syringe, which is the recommended dosage for adults, but I kept refilling the tiny 3 cc syringe for her. Shortly, it seemed that the sugar had affected the little guys, and all three were standing and more alert. I told Katherine to try and help them nurse while I ran into the house and tried to find the little goat coats we'd made for kids a couple years earlier.

Although unable to find anything for the kids to wear, I did find the remains of the sweatshirt that I had used to make the kid

This is a picture of Sadie in labor. Whenever a doe starts pushing, I turn on the heat lamp if it's cold outside. That way I don't have to think about it after the kid is born. We have a mix of clear and red heat lamp bulbs. I prefer the red heat lamps because I think it's less disruptive to the goats trying to sleep at night. However, red heat lamps do cause challenges when it comes to photographing births and kids.

coats, and it occurred to me that it might fit Sherri, so I took it back outside and put it on her. She didn't object to being dressed, and eventually her shivering diminished. Then I noticed steam rising off the kids under the heat lamp, and when I felt them, the tips of their hair were frozen. Obviously, "dry" has a different definition at −4°F, so I grabbed the hair dryer and started blowing on their coats again. I had to hold it a couple of inches from them because the hot air would not travel any farther than that before it was cold. At some point, we also put a heating pad out there for them to lie on.

Two hours after they were born, we were eventually convinced they were dry and warm. Two of the three kids had nursed, and Sherri had passed her placenta, so we felt we could go inside for lunch. Katherine went out to check on them and discovered one of the boy's ears felt frozen again, so she warmed them up until they felt flexible. I'm not sure what to do about the ears. It's −3°F now and supposed to get colder tonight. It feels like a continuing battle until the temperatures go up again in a few days.

Many goat owners are worried about breech or posterior births, but they are not necessarily dangerous. A breech kid can be a little harder for a doe to birth because the butt is bigger than the pointy nose and head, but unless the kid is especially large, a doe can push out a breech kid. The first kid usually takes the longest for a doe to push out, and if it's breech, it can feel like it's taking forever. As long as the kid's body is still entirely inside the doe it's fine. Once the hips were out, Sherri pushed the kids out quickly, so I didn't need to do anything other than catch and dry. However, if she had decided to take a break after pushing out the back end, I would have pulled the rest of the kid out in case the umbilical cord was broken, which would have meant the kid's oxygen supply was cut off while its head was still inside the doe.

When temperatures are below zero Fahrenheit, even the easiest births are a big challenge. When Sherri gave birth to the three bucks, it was our first experience with a doe giving birth at that temperature, so we didn't realize that frostbitten ears were a possibility, although we knew that the kids could get hypothermia and die very quickly. Does will lick their kids when they are born, but a doe has a very small wet tongue, which does nothing to truly dry a very wet kid. You need one or two large towels for each kid plus a blow-dryer and heat lamp when it is that cold. You need to be especially mindful of the ears, tails, and feet, as they can freeze and will then fall off a week or two later. If a kid loses all or part of its ears or tail, it will just look funny, but if it loses its feet, you may need to euthanize the kid because it will be disabled.

I no longer give Nutri-Drench to kids that are slow to get started after birth. Instead I milk the mom and give the colostrum to the kids, using either a syringe or a bottle. Colostrum is truly liquid gold and gives the kids so much more than just sugar.

❖ ❖ ❖

The next story was our second experience with kidding below zero. We have now had seven or eight does give birth in below-zero temperatures, and it never gets easier or less stressful.

B·L·O·G
MON
FEB 14
2011

Jo's quads

February 9—Jo's tail ligaments were rock solid, so we were all thinking she would be the third or fourth to kid. Charlotte's ligaments had been soft as a rubber band for days, so we were carefully watching her, thinking that she could drop kids at any moment. I even spent one night in the barn when it was really cold because I didn't want the kids to get hypothermia. We finally found the baby monitor and got it hooked up so I could sleep in the house.

Wednesday evening, one of the goats was bellowing over the monitor for quite some time. Although it didn't sound like she was in the midst of actually giving birth, I sent Katherine to check on her. She came back with an inconclusive report on Jo. So, before I went to bed at 11:00, I decided to go check on her myself, even though she had been quiet for the last couple of hours. When I saw her and checked her tail ligaments, I thought, *Yes, she could go at any time.* So, I put her into a kidding pen with clean straw and bucket of warm water and went back inside the house.

I was brushing my teeth, and Jo continued to make little "meh, meh, meh" sounds. As I was just about to get undressed for bed, Mike said, "I think I hear a kid." I listened, but I didn't hear anything. He insisted. "Don't you hear it?" Knowing that I would forever feel guilty if I didn't go check on Jo, I headed downstairs. As I was putting on my insulated overalls—because it was 8 BELOW ZERO outside—Mike yelled down the stairs, "That's a newborn sound!" He sounded really convinced, so I hurried outside. He said he would listen at the baby monitor, in case I needed help.

I walked into the kidding barn and saw Jo lying right where she had plopped down when I had left her fifteen minutes earlier. At her back end were three babies in a puddle of amniotic fluid. I started screaming a bunch of stuff that I didn't even remember five minutes later. It really didn't matter what I was screaming. I knew if Mike heard me, he'd come running, and I knew I needed all the help I could get. If we didn't get the kids dry and warm quickly, their ears could freeze, or worse, they would die of hypothermia. I ran to the office and grabbed the blow-dryer and heating pad. When I got to the pen, I realized I needed towels, so I dropped everything and ran back to the office for the towels, running smack into the handle of a wheelbarrow in the darkness.

Having no idea which kid had been born first, I spread out a towel, grabbed two kids and wrapped them up in it. I toweled off the third one and began blow-drying it. As soon as Mike arrived, I told him to start toweling off one of the other kids. We didn't even stop to check the sex of the kids until we'd been working on drying them for at least five minutes. Jo was making the most horrible noise. She sounded like she was dying, and she was shivering badly. We draped a towel over her back to help warm her. As soon as Katherine arrived, I told her to get the Nutri-Drench, which is a molasses and vitamin concoction that gives goats a little extra energy. I squirted some into Jo's mouth, and she continued making the pathetically weak bleating sound. About ten minutes later, she pushed out a fourth kid—a doe. All of the kids were at least 3 pounds, so I wasn't worried about any of them.

It took two hours to get the kids dry, and by then, the placenta had passed. We helped the kids get started with nursing, and then, shortly after 2:00 a.m., we went to bed.

Most of our does eat their placenta as soon as it comes out. This is a great example of a normal, healthy placenta. The cotyledons should look like dark purple, almost black, prunes. If a kid is born dead or mummified, some of the cotyledons may be lighter purple or even gray. It was about half gone by the time I grabbed my camera to get the photo.

This is a perfect example of why I tell people that the number one, most important thing they need to have in preparation for kidding is a baby monitor. Yes, I'm talking about a $20 or $30 monitor that people use with their human babies. People worry about things going wrong, but sometimes things go right at the horribly wrong time, as in this case. I have heard so many stories of goat owners who left a doe like Jo and returned a couple of hours later to find dead kids. That is exactly what would have happened in this case if we had not had a baby monitor that alerted us to the fact that she had given birth.

I didn't usually mention the placenta in most of my birth journals because it has always been my habit to let the moms eat it, if they want to. In nature, mothers usually eat the placenta, and goats are no exception. It is highly nutritious. First fresheners are the least likely to eat it, but most mature does love it. Some goat owners don't want to let their does eat it because they are worried about choking. While it is true that they could choke on it, just as they could choke on anything they try to eat, it's not common and has never happened with any of our goats. If the doe is showing no interest in it after thirty minutes or so, we feed it to our livestock guardian dog.

When Victoria was pushing, she was lying on her side with her legs pushed straight out in front of her body while she screamed. Rather than craning her neck, though, she turned around and looked towards her stomach or back end. Being a first freshener, she had no idea why that part of her body was in so much pain. Experienced does may "talk" softly to the kids during the later stages of labor rather than screaming, but they may look very similar to this.

Victoria's twins

B·L·O·G
WED
MAR 4
2015

The day before Victoria kidded, I realized that I had not appreciated my senior does enough. You see, when a goat has given birth before, they are usually very stoic until the last minute or two before the kid is actually born. Victoria was a first freshener and completely freaked out by every little contraction. Sunday night after Cicada gave birth, Victoria was screaming her head off, sounding quite unhappy. I went running out to the barn a few times, only to find her standing there screaming as if to simply express her displeasure. She was standing like a normal goat and just screaming. She wasn't pushing or lying down or pawing the ground or doing anything that a goat normally does when in labor. Even though she is not a friendly goat, she was very unhappy and would scream more whenever I left her. So I wound up staying with her for three hours! And then she was quiet. I went inside, went to bed, and went to sleep.

Monday morning shortly after Mike left for work, she started screaming again. I decided to take a book out to the barn office and read, because it was obvious that she was not going to stop screaming anytime soon. Even though it was a really dreadful scream, her body language just

Some goat owners get very worried if they see more than one bubble emerging at once. Although it can mean that two kids are trying to be born at the same time, it may mean nothing. In Victoria's birth it meant so little that I didn't even mention it in the original journal I wrote about the birth.

The kid's leg is upside down. The dew claw should not be facing Victoria's tail. I realized it was a front leg rather than a hind leg because there is no hock.

looked like she was mad about something, so I wasn't taking it very seriously.

An hour and a half after I went out there, she finally got serious, and it was obvious she was actually pushing. She'd lie down, throw her head back, curl her tail over her back, and stretch her legs out in front of her as she screamed. She also quit eating. And then things got interesting.

A hoof was sticking out, but it was upside down. That meant that it was a front hoof from a kid that was upside down, or it was a back hoof. A back hoof would have been the preferred presentation. It would actually be pretty easy for Victoria. However, after ten minutes of pushing, the foot was sticking out about 3 inches, and there was no sign of progress. I ran my finger along the leg and bumped into a nose and mouth. The books tell you that in this situation, you should reach in and flip the kid over. I was home alone, and the odds of Victoria standing there while I did that were somewhere between zero and never. The idea of doing that also worried me because of the risk of tearing the uterus. I went looking for some disposable gloves and some iodine, but I wasn't entirely sure what I'd do when I actually had them. In the meantime, Victoria

Although it can be more challenging for a doe to give birth to a kid that's upside down, Blanche also gave birth to this upside down kid unassisted.

kept pushing. About fifteen minutes later, the kid was born. The head actually came out sideways, which I don't recall ever seeing before, and the body came out with the kid's belly facing Victoria's tail, which is basically upside down. The little doeling was in great shape, and as Victoria and I started to clean her up, I noticed another upside down hoof sticking out of Victoria's back end.

Victoria and her two rather large doelings shortly after birth.

"Seriously?" I asked Victoria. "Another one?" I ran my finger along the leg, and when I came to a joint, I bent it. It bent in the direction of the top of the hoof, and that meant it was a hind leg, which should be a much easier birth than the kid that she just delivered. And a couple minutes later, the second doeling was born quickly.

Victoria has been an excellent mom from the very beginning. I think she heard that I'm planning to seriously reduce our goat herd this year, and she wants to be sure that she makes the cut and gets to stay here. After all, it was pretty impressive for her to give birth to doelings that weighed 3 pounds, 4 ounces, and 3 pounds, 9 ounces, especially since one of them was upside down, which is never easy, even with smaller kids.

Since my daughters grew up and left home, I've attended a lot of births by myself. That means that when I see an abnormal presentation like this, I usually sit and think about how I'm going to help. Unless a doe is exhausted, she is unlikely to just lie there while you stick your hand inside of her. While I'm thinking, the doe usually winds up doing it all herself, even though a lot of books or experts would say that it is impossible. Having seen what goats can do, I realize that probably there were times in earlier years that I helped when it was not really needed. I have learned that in most cases I have plenty of time to watch and think, and this has helped me to remain calm now. I've discovered that the more births I attend, the less likely I am to rush in and try to help.

Kidding at 17 below zero

·B·L·O·G·
THU
FEB 6
2014

For days the weather prediction for Sunday night was a low of −13°F, but then on Saturday, they were suddenly saying −3°F, which was certainly music to my ears, because Agnes had not kidded yet. In jest, I posted the following status on the Antiquity Oaks Facebook page on Sunday morning:

> Dear Sweet Wonderful Agnes (you know, you've always been my favorite)—PLEASE give birth this afternoon at 2:00 p.m., when the high temp will hit 17. If you can't manage to give birth by sundown, then PLEASE hold on to those kids until tomorrow afternoon, when the high will again be 17. Do NOT under any circumstances give birth in the middle of the night, when the temperature is supposed to be 3 below zero! Your cooperation in this matter will be SO much appreciated! Hugs!!!

A couple of people responded that I had just guaranteed a middle-of-the-night kidding. And they were right. We have a video monitor over the kidding pens so we can watch and hear what's happening in the kidding barn from the warmth and comfort of our bed in the house. At 2:59 a.m. I was awakened by that familiar scream of a goat giving birth. And this is Agnes, a Sherri granddaughter. If you've been around this blog for very long, you know that Sherri and her daughters and granddaughters don't make a sound until the kid is actually coming out, and they give birth very fast.

Mike and I jumped out of bed and pulled on multiple layers of clothes as quickly as we could. I actually keep all of my layers together—turtleneck inside the sweatshirt and long underwear inside the sweatpants—laid out in just the right position, right in front of the toilet, so that I jump out of bed and go sit on the toilet as I'm changing clothes. Multitasking! Every minute counts when you have a 3-pound, soaking wet kid in sub-zero temperatures.

Mike made it to the barn first and began toweling off the little doeling that had already been born. Only a moment after I arrived, the second kid came flying into this world. I turned on the blow-dryer and spent the next hour and a half blow-drying the two little doelings. After returning to the house, Mike checked weather.com, and they said the temperature was −17°F, which would explain why it took so long to dry the kids! We've had births in the single digits below zero before, and it took only about an hour to blow-dry the kids, but the colder it is, the harder it is to dry them. I had to hold the blow-dryer within a couple of inches of the kids, and the only part that was being dried was the exact spot that I had targeted at any moment. Every time I switched kids, it was quite obvious where all the little cold, crunchy, frozen bits were.

Although one of the kids was pretty quick to start nursing, the other one seemed clueless. We were finishing up the drying, and she still had not tried to nurse. I put my finger in her mouth to discover that it was surprisingly cool and the kid had almost no sucking reflex. I asked Mike to go inside and get a bucket to milk Agnes and then a bottle for the kid. I wrapped the little doeling in the heating pad in my lap until Mike returned with the bucket and bottle. Of course, as soon as the colostrum hit the bucket in that temperature, it was cold, so he had to get a second bucket with hot water in it. We placed the bottle in it to warm up the colostrum. Finally, I was able to get some warm colostrum into the little doe, and within five minutes, she jumped out of my lap, ran up to Mom and started looking for the teat!

Mission accomplished! It was past 5:00 a.m. when we finally got back into the house. I was frozen and sore, so I got my own heating pad and lay on it for the next two hours, moving it from my feet to my back and neck and everywhere in between. Mike got half an hour of sleep, and then at 7:00, he had to get up to do chores before going to work at the college where he teaches.

I wanted to include this story because it illustrates that challenges come in so many shapes and sizes. It's not just about the kids being in the right position. Weather during kidding is, however, something that you have some degree of control over. Although you can't control the weather on any particular day, you can certainly control the likelihood of a sub-zero birth by breeding your does when that is impossible or at least unlikely. Today I breed for March to May kids.

Commenters on my blog following these insanely cold kiddings would always ask why we were having kids in such terrible weather. No, we didn't have a bunch of accidental breedings. Winter kiddings were part of our strategy to overcome a problem with dewormer resistance. I started raising goats in 2002, when the standard advice for parasites was to deworm the whole herd at one time, deworm and then move to clean pasture, deworm on a schedule, deworm all does after kidding, and rotate dewormers. Those recommendations were based on assumptions about what would work. Research done in the past ten to fifteen years has shown that those practices lead to dewormer resistance because you are repeatedly exposing the worms to dewormers. Worms that survive the deworming have baby worms that are resistant to the dewormers. That meant that after raising goats for a few years none of the dewormers worked, and goats were dying from worms.

It is true that the stress of kidding can cause a doe to have an increase in her parasite load, but it is not inevitable, and there are strategies you can use to reduce the risk, such as kidding when parasites are at their lowest level on your pasture. For us, that meant kidding in the dead of winter when there were no parasite larvae on the grass. For others that might mean kidding when it's especially hot and dry. So, although the doe's egg count might increase, it won't negatively affect her because she is not increasing the load even more by ingesting larvae in the pasture.

I discuss internal parasite control in goats as well as dewormer resistance in more depth on my Thrifty Homesteader website, thriftyhome steader.com, and in my book, *Raising Goats Naturally*.

3

NOT SO
NORMAL BIRTHS

A multitude of things can go wrong during the birthing process. Luckily, they don't usually. The following stories are of complications that have happened only once on our farm and probably won't happen again for a very long time. However, there are no guarantees. It is very common to make a mistake, but always remember that mistakes are inevitable. And it is common to feel guilty when there is a bad outcome, but sometimes nothing anyone could have done would have resulted in a better outcome.

TUE
SEP 8
2009

Always more to learn: Giselle's birth

I was sitting in bed Monday morning, enjoying one of Mike's homemade croissants while checking email, when Katherine came running in babbling incoherently. I blurted, "Giselle!" and spewed masticated croissant all over my laptop.

"Yes! There's something huge sticking out. It's either a side or a butt or something. It's really hard."

As I jumped up and started to change clothes, I realized that Katherine would know a butt if she saw one. We've had plenty of breech births, and they're usually not a big deal. When I got to the barn, I was surprised by a large, flat thing that was presenting itself—something that looked completely unfamiliar. I felt all over it, trying to figure out what I was feeling. I closed my eyes and tried to visualize the bones I felt. It didn't make sense. I tried to talk myself through it. "Okay, there's a bone going this way that feels like a leg bone, but there's a bone going this way that feels like a leg bone. This feels like a spine." But none of the parts were in the right places, and everything looked completely smooth.

In these situations, everyone says "Just push the baby back inside the uterus, turn it around to a good position, and pull." The way they say "Just push the baby back inside" makes it sound so easy. As I tried to push it back inside, I found it extremely difficult, both physically and mentally. Giselle didn't want the baby back inside, and it was completely counter-intuitive for me to try to push the baby back inside. I pushed, but I couldn't do it. I tried to find something to grab onto and pull, and somehow I got the baby out. The whole ordeal was more horrible than I can explain right now, and trust me when I say that none among you (except the hardiest farm women) really want the details. Next time, I will push with every ounce of energy I have to get the baby back inside, turn it to a proper position, and then pull it out.

He was dead. I laid him on a towel and covered him up. I don't remember what Katherine and I were saying as we were sitting there watching Giselle, but a couple of minutes later, without a single push or sound, a kid literally fell onto the straw. As I picked him up and started to wipe off his nose, a second nose started coming out. I tossed a towel at Katherine and told her to catch the kid and clean it up, since I hadn't even cleaned the nose off the first one yet. When we put the babies in front of Giselle, she started licking them, but she didn't attempt to get up.

About half an hour after the kids had been born, Giselle still was not standing. I realized there wasn't a water bucket in the stall, which had just been cleaned for this very purpose. When Katherine brought the bucket of water, I told her to put it far enough away that Giselle would have to get up to reach it. One of her kids was wobbling around looking for his first meal, which would be impossible to obtain from a mother who was lying down. Giselle got up and took a couple of steps to reach the bucket. She walked like a drunken sailor. *Not good,* I thought. I put the little buck under his mama so he could start nursing. Giselle's back end swayed to one side. I put my hand up to stop her from toppling. After a few minutes, her kid stopped nursing, and she plopped down on her butt unintentionally.

I gave her some Nutri-Drench and a shot of BoSe, a selenium supplement. There was little change throughout the day on Monday. She always seemed to be lying down, and she was wobbly when she was standing. She stood enough that the kids were able to get full tummies. She passed her placenta, but there appeared to be an umbilical cord hanging out. Another breeder and a vet had once told me that was a sign of a retained kid.

I straddled her body, bent over, and put my hands under her belly right in front of her udder. I lifted, felt nothing hard, put her down, moved my hands slightly, lifted again, felt nothing and repeated the exercise a few more times. I thought I felt something hard, but was it a kid's bones or the doe's ligaments? Do does have ligaments there? I went to the stall next door and checked two of my milkers. Their bellies felt completely soft. I checked Giselle internally for another kid but found nothing. I gave her a shot of antibiotics and hoped that I had checked her thoroughly.

This morning, the cord is still hanging out. Her belly looks deflated. The straw is stained yellow in several areas around the stall. Perhaps she just had a lot of fluid in her uterus that made her look so big yesterday after she'd given birth. I've had a partially retained placenta before; it didn't look like an umbilical cord. I have a thought haunting me; I know someone who lost a doe because of a retained kid.

After seven years and more than a hundred does kidding, I feel like I should know more than I do. I should have fewer questions. I should feel more confident, but instead of confidence, I only have hope.

The good news was that Giselle did not have a retained kid. It was merely a piece of retained placenta, which passed on its own a couple of days later. It's likely Giselle ate most of the placenta because she got tired of dragging around the part that had already passed. The placenta is a very large membrane, and about two feet of it can be hanging out of a doe before the whole thing detaches and falls out. I have seen goats reach back and grab their placenta with their mouth and give it a yank or start eating it. Now if I have a doe that is yanking at the placenta before it is fully detached, I tie it up into knots so that she can't reach it.

At the time I wrote this story, I couldn't provide the details of actually getting the kid out because I was too upset and feeling too guilty. After speaking to a pathologist, I had a better understanding of what happened and realized that the kid had been dead for quite some time before Giselle went into labor. I've also learned since then that when you have really crazy positions, like ribs first, it's often because the kid is already dead. Live kids almost always get themselves into the nose-first diving position to be born.

Because of this birth, one of my cardinal rules of assisting is "Never pull unless you know what position the kid is in and you know what you are pulling!" I should have figured out exactly how the kid was positioned before I started to pull on the front leg that I found. As I was pulling, the skin on the leg started to tear. It was the only time I have ever started crying at a birth. But the tearing skin also indicated that the kid had been dead for at least a few days, according to the pathologist, because the skin was very thin and easily torn.

I know someone who actually pulled the leg off a kid that had been dead for even longer and was in a more advanced stage of decay. Unfortunately, if a kid has been dead in utero for a few days or longer, there isn't anything you can do for the kid, but it usually helps to know that live kids don't normally fall apart when you are pulling them out of a doe. When a mummified kid or one that has been dead for a few days or longer is born, you might consider administering antibiotics or herbs that support the doe's immune system. At a minimum, you need to monitor her temperature to be sure that she doesn't get a uterine infection.

Our most puzzling birth ever

B·L·O·G
WED
APR 18
2018

I'm writing this story almost two years after it happened, but the whole thing was videotaped, so I am actually able to recount it with much more detail than the other stories. This doe and her sister were born on my farm and had been brought back for breeding. Then the owner decided he didn't want goats any longer, so they stayed here, and ultimately I'm glad they did, especially Ellie. This was one of my most challenging births ever, even though I'd attended about six hundred births previously, and I can't imagine how a first-time goat owner would have responded to this.

My husband and I were awakened at 4:50 in the morning because we heard Ellie making pushing noises over the video monitor. We threw on our clothes and ran out to the barn to find that she was indeed pushing, and there was a small bubble emerging with each push. It would open her vulva about 2 inches. Then the contraction would end, and the bubble would mostly disappear. After six or seven contractions that were all exactly the same I knew something was wrong, so I decided to do an internal exam.

"This is a very weird presentation," I said. "I don't really know what I'm feeling. I think there's two kids. There's either two kids or one is in a U-shape. That's why she's pushing but not making any progress." Then I realized I was feeling the top of a kid's head and I recognized the feeling of an ear. An ears-first presentation is very difficult for goats. It's usually necessary to push the head farther back into the uterus so that you can flip the chin up and then grab a leg or two and pull.

And that's exactly what I did. I was able to grab both legs and pull, but just as I was nearing the exit, the legs slipped out of my hand. One hoof was left sticking out of her vulva, so I decided to give both of us a break at that point. Ellie closed her eyes with ears pointed back and was completely silent. Then a minute later, another contraction started, and a tiny kid shot out with very little effort. The only reason she had been having a hard time was because the kid had originally been presenting with the top of his head pointing towards the exit.

I dried off the kid and laid him down on the towel in front of Ellie. She sniffed him and then turned away. I realized that the kid was not lifting his head or making any noise, so I picked him back up and started to rub his chest briskly. I really wished I'd had my bulb syringe because I worried that he might have too much mucus in his airway. Unfortunately, it had been left in the barn a couple of days earlier and fallen into a kidding pen, where I later found it chewed up and worthless. So, the best I could do was hold him upside down and hope that gravity could help while I continued to rub him briskly. He let out a few cries, which were weak but encouraging. Because it was freezing and he didn't have much energy, I laid him on a heating pad in front of Ellie's face.

About fifteen minutes after the first kid was born, he was starting to scream with a little more conviction, and Ellie lay down and gave the tiniest push. I saw two hooves sticking out, which was initially encouraging until I realized it was one normal presentation and one upside-down hoof. That meant that two kids were trying to be born at once. One was presenting normally, and the other was either upside down or posterior. Then Ellie closed her eyes and looked like she was going to go to sleep.

I inserted two fingers into her vagina and felt a head that belonged to the correct hoof. I decided to push back the upside-down hoof with my thumb, and as I was doing that, a kid shot out so forcefully past my hand that I screamed in surprise. The kid landed on the straw, and I grabbed it and began to dry it. After drying it for a few seconds, I realized it was not moving. I put my fingers on its chest and felt nothing. No heartbeat. It was dead. Then I grabbed some hairs between my fingers and easily pulled them out. That meant the kid had been dead for at least a few days, according to a pathologist that I talked to after Giselle's birth, when I delivered a dead kid whose skin I accidentally tore when assisting in birthing it.

Only a minute and twenty seconds after the dead kid was born, another one shot out of Ellie without her making a sound. (One advantage of videotaping a birth is that you know exactly how much time passes.) I picked it up off the straw and started to wipe off the nose so that it could breathe, being skeptical about it being alive. As soon as the nose was clear, I checked for a heartbeat. There was none. This kid was also dead. I was also able to easily pull some hair from this kid, although not quite as much, so it had probably died a little more recently than the other, but still a few days ago.

Ellie looked like she had no energy left, and there were no membranes hanging out of her vagina. Normally when a doe has given birth to the last kid, there are at least some umbilical cords hanging out, if not a lot of membranes and the start of the placenta. So I decided to insert a finger to see if I felt another kid hanging out at the exit. I felt a tennis-ball-sized bubble, which she pushed out. Then when I inserted a finger again, I felt the top of a kid's head. It was another ear-first presentation! In almost six hundred births, we had had only two ear-first presentations before Ellie went into labor, and now we were about to experience our fourth!

This kid was also very small, and I only had to push it back slightly to be able to flip up the chin so that it would be easier to birth. The front feet slipped out as soon as I had the chin up, so I grabbed the feet and gave a tug, and the kid easily slipped out into the world. This one was making noise as soon as it was born, letting us know that it was in great shape.

Ellie still didn't have much hanging out in the way of membranes, and since all of the kids were weighing in at less than 2 pounds, and two were born dead, I was worried that there might be another kid or two. That could explain the small size and the fact that two had died in utero before the end of pregnancy. I did another vaginal check and couldn't find more kids, so we waited for the placenta. I had both kids lying on a towel in front of Ellie's face, but she wanted nothing to do with them. That's really not much of a surprise for a first freshener. They have no idea what's happening when they're in labor, and when a birth goes as terribly wrong as this one did, the doe can be so traumatized that she is not interested in the kids at all.

When Ellie passed her placenta, we began to unravel the mystery behind the terribly small kids, the two dead kids, and the challenging birth. A normal, healthy placenta looks like a giant sheet of clear membranes with a bunch of dark purple prunes attached to it. Those prunes are cotyledons, and that's where the exchange of nutrients takes place between the mother and the kids. About half of the cotyledons on Ellie's placenta were a lighter pink color and looked rougher than normal. So, half of her placenta was no longer working, and the two kids whose umbilical cords were attached to that section had died.

But why? Luckily, the previous owner was willing to help get this figured out. I ultimately learned that he had not purchased the brand of goat mineral I had recommended. Instead, he had bought a brand I have never heard of, and when I searched online for information about it, I learned that it was little more than salt with no meaningful level of minerals, so Ellie was probably deficient in multiple minerals. Then it became obvious to me that she was deficient in copper because she had no hair on the bridge of her nose and a lot of her bright red coat had faded to cream. I had not even thought about her needing additional copper supplementation because she had been living in the city, where there were no copper antagonists in the water. Here on the farm we have high sulfur and iron in our well water, which bind with the copper, meaning that our goats need more copper to make up for that. Copper is stored in the liver, though, and deficiency takes months, so it never occurred to me that this city goat would be at risk of copper deficiency—or any other deficiency for that matter.

But there was even more. As it turns out, they had been taking Ellie and her sister for walks in the city, and the goats were snatching up cigarette butts and eating them off the sidewalks. I, essentially, had a couple of goats who were the equivalent of chain smokers when they got pregnant. Cigarette butts are high in a variety of chemicals, including nicotine, which causes low birth weight and impaired placental function in human babies. Ellie's sister had given birth a few days earlier to kids that were also very small, but they all survived, probably because she was carrying only three kids rather than four, so her body was better able to allocate the nutrients between them.

All of the other kids that year and for many years on our farm have been closer to 3 or 4 pounds, while Ellie had given birth to four kids under 2 pounds, and her sister gave birth to kids that were barely 2 pounds and a few ounces. Many years ago, when our goats were dealing with nutritional deficiencies, we saw smaller kids, but once we figured out the cause, we had normal-sized, healthy kids. I believe that if you get the nutrition right, everything else falls into place for your goats' health. If you get the nutrition wrong, you can't do anything to compensate. You will simply have a lot of problems. I am reminded of something I heard an experienced holistic vet say—"I've seen a lot of people go broke over the years successfully treating sick animals." I was almost one of those people. I spent a small fortune on veterinary bills to save goats in the early years before I realized we had issues with nutritional deficiencies. After getting the nutrition figured out, the vet bills have been zero most years.

Even though Ellie had a terrible birth and rejected her kids, we started milking her immediately, and she turned into such an excellent milk goat, we decided to keep her. The following year she gave birth to healthy twins. One weighed in at 4 pounds while the other was 3 pounds, 9 ounces.

Lizzie was pushing hard and screaming, with her head and ears back and her legs pushed straight out in front of her.

Lizzie's triplets

B·L·O·G
TUE
MAY 18
2010

Margaret arrived home from college on Thursday when Katherine and I were gone. As we pulled into the driveway upon our arrival back home, we saw Margaret running into the

One reason we didn't get worried sooner is because Lizzie was fine between contractions, even licking my daughter's face.

Between contractions she was alert and didn't seem tired.

kidding barn. Katherine stopped the car at the end of the driveway, where Jonathan was mowing the yard. As I rolled down the window, he called, "Lizzie is in labor!"

Katherine pulled forward and stopped the car. We ran into the house. "I have to change clothes," I hollered as I was running upstairs. I was not going to attend a goat birth wearing beige pants. Jonathan yelled that Margaret needed towels.

When I entered the kidding barn a few minutes later, wearing farm clothes and carrying a stack of goat towels, Margaret and I started talking about the end of her semester and what had been happening on the farm lately. Katherine gave the bottle brats their afternoon bottle and went back to the house to get her new camera to take pictures of the birth.

The three of us sat in the pen with Lizzie as she screamed during contractions, sometimes nearly rolling onto her back. She would lean against one of us, then stand up and move to another person for the next contraction.

"Shouldn't we have seen some progress by now?" I don't remember who was the first to voice a concern, but we all agreed that we should be seeing something by now. Margaret looked at her watch. Half an hour had passed since Lizzie started screaming through contractions, and her back end was completely unchanged. There was no bulging, no thinning skin, and definitely no sign of a nose or hoof. One of the girls said, "Someone needs to stick a finger in there and see what's up." I remember a time when the two of them would fight over whose turn it was to deliver a goat's babies. Now it was obvious that "someone" meant me. "Okay, fine, I'll go get gloves."

I squirted iodine on the glove and attempted to figure out what was where. The first thing I felt was a joint with two thin bones attached to it, but then it was gone. "That felt like a hock, but it can't be a hock." Okay, yes, I know it could have been a hock, but I really did not want it to be a hock. It is amazing how much a nose and a goat's butt can feel similar when you can't see them. I convinced myself it was a nose, and since it was still 3 inches inside the goat, I had nothing to grab easily, so I decided to wait a few more minutes.

The girls and I continued talking, and twenty minutes later someone commented that there was still no visible change, although Lizzie was getting tired. Her body felt hot and sweaty. I checked again to find that the kid was in exactly the same place as it had been twenty minutes earlier. I realized that I'd have to put my whole hand inside her to be able to grab the kid and pull it out, so I went to the other barn to get the kidding box.

The kidding box is a plastic toolbox that we rarely need. It contains the

Lizzie's tail is on the left side of the photo, and her feet are towards the bottom of the photo. The first kid was presenting butt first and upside down. I pulled him out by grabbing the uppermost part of his leg where it meets the pelvis. If he had been too big to fit, I could have pushed him back into the uterus so that I could grab his hind legs and pull him out back feet first.

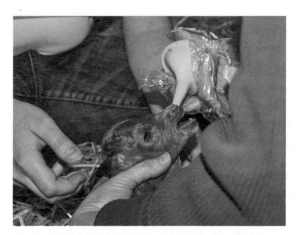

To suction kids back then we were still using the bulb syringe that we had used when our human kids were infants. You can see that I still had the plastic shoulder-length OB sleeve on my right hand and arm.

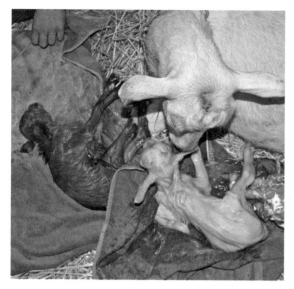

After the second kid was born, I placed him in front of Lizzie's face so she could lick him.

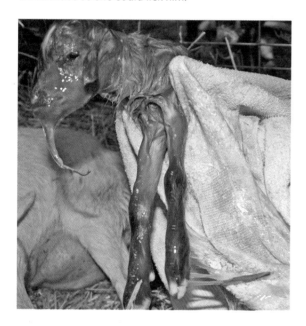

A lot of people are surprised by the amount of mucus on kids when they're born. This is not unusual.

emergency stuff—the kid puller, the shoulder-length gloves, the bulb syringe, and stuff like that. The shoulder-length gloves have been in there for six or seven years, and Margaret used one five years ago. That is how seldom these goats have problems with kidding.

As I pulled the giant glove over my hand and up to my shoulder, I said, "These things are made for men pulling calves." And one of the girls said, "They're made for fat men." I laughed.

Saying that I put my hand inside her and pulled the kid out makes it sound so easy. It wasn't easy—not for me and certainly not for Lizzie. In addition to being breech, the kid was also upside down, meaning that it was butt first and instead of the kid's spine being against the mama's spine, it was the other way around. Once the kid was out, I started wiping off the mucus. I felt no movement beneath my hands. Someone asked, "Is it alive?"

I shook my head and said, "I don't think so." Lizzie was already pushing to birth another kid, so I handed the kid to Margaret and said, "Here, keep working on him. Rub him like this." I demonstrated by rubbing briskly up and down his body, even though Margaret has delivered plenty of goats herself and knew what to do.

"He's alive! I felt him move!" Margaret said. "He's trying to breathe, but he's mucusy. Where's the sucker thingy?"

Lizzie's contraction had ended, so I pulled the bulb syringe out of the kidding box and started to suction the kid's throat and nose. He was very mucusy. After I'd suctioned his nostrils and throat several times each, Lizzie started another contraction, so I handed the bulb syringe to Margaret and turned back to Lizzie just in time to catch the second baby. As he wiggled and kicked and bobbed his head up and down, I realized the first kid was still quite weak. As soon as I had the nose clean, I placed the second kid next to Lizzie, so she could start cleaning him up.

A couple of minutes later, the third kid flew out like a torpedo, and the moment I had cleaned her nose and mouth, she was kicking and screaming. "That's what I like to hear!" I laughed. Within ten minutes, she was hoisting her back end up in the air and trying to stand. Kid number two followed suit. In little time, the two of them were wobbling around and bopping on Lizzie's chin, looking for dinner. When Lizzie stood, the little doe was nursing before we even realized she was trying.

We gave the first kid 3 ccs of Nutri-Drench about thirty minutes after he was born because he still was not holding up his head. An hour after he was born, the first kid finally stood. We went into the house for dinner, and when we came back afterwards, he was nursing. The next morning, he was still doing great.

When kids start nursing, they might wind up with blood from the placenta on their head or body.

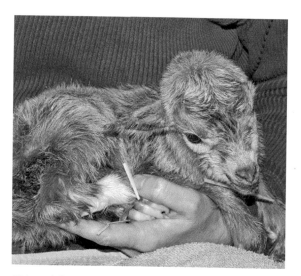

Although it took more work than usual, with a lot of brisk rubbing and suctioning, the first kid did eventually perk up.

This was the first upside-down breech kid we'd ever had born on the farm, which is what made it more challenging than the other breech kids we'd had. As you've read in other stories, many breech kids can be born with no assistance. But when the kid is upside down, the curvature of the kid in relationship to the doe's pelvis changes, and it is more challenging for the doe to push out the kid unless it's very small.

When faced with this type of presentation, some recommend flipping the kid to a nose-first presentation, but I was told many years ago by an experienced breeder that pulling hind legs is actually easier for a doe to birth than a nose-first presentation, and I'd agree with her. It's also much easier to simply find the hind legs and pull than it is to try to turn a kid in utero to a nose-first presentation.

Alex's final kidding

SUN MAY 22 2016

Before I tell you this story, I have to tell you that we've had twelve does give birth so far this year, and ten of them were textbook perfect. Now I'm really wishing I had been able to find the time to write about those joyful, beautiful births.

On April 18, I happened to look out the window by my desk, and I saw Alex lying on the ground in an unmistakable position. Her legs were outstretched, her neck was craned, and her ears were pointed back. She was in labor and pushing. I jumped up, grabbed a stack of goat towels that just happened to be next to the front door and ran to the pasture. It was a beautiful spring day to have a goat give birth in the pasture.

As soon as I opened the gate, Lucy, the Great Pyrenees, ran towards me, and as I approached Alex, the dog's exuberance caused the goat to stand up and start walking away. I put Lucy into another pasture and went back to Alex. Of course, I was wearing a pair of my town jeans, which means that they didn't have any farm stains on them, such as manure, dirt, or blood.

Beauty, the Jersey cow, also happened to be in the pasture, and she decided to come say hi. She has recently decided that I'm her new scratching post, and she started to rub her head against me. I called Sarah, who was in the barn working on a building project, and she came out and moved Beauty to the next pasture. Then she brought a flake of straw so that I could scatter it on the ground because Alex had chosen to lie down in a spot with more dirt than grass.

The first kid was very slow to come out, partly because Alex wasn't pushing very hard and partly because he was quite large. I'd later discover that he was 4 pounds, 7 ounces, which is big for a Nigerian, where 3 pounds is more average. Because her contractions seemed to be really wimpy and the kid was making little to no progress from one contraction to the next, I pulled on his legs to help bring him into the world.

I could see that the second kid was even bigger when the first hoof emerged. I kept comparing it to the hoof of the kid that had already been born and was worried. As the nose and white head began

Kids usually have little to no blood on them when born. This kid's head was covered with blood when born, and this is what it looked like after I had wiped it off with a towel.

to come out, it was easy to see that there was an unusual amount of blood. Turns out he weighed 5 pounds, 2 ounces, setting a new record for largest Nigerian kid born on the farm. Because he was also progressing excruciatingly slowly, I wound up pulling on his front legs to help Alex when she pushed, and it was such a tight squeeze that Alex was even pulled across the grass a little.

After the kids were born, I started to worry even more because Alex completely ignored them. She's an experienced mom and has always been wonderful with her kids. We took her and the bucklings into the barn, and not only would she not let them nurse, but she wouldn't let me milk her. This is a doe who had finished a three-year lactation two months ago, so she's also a very experienced milker! She just wanted to lie in the stall with her ears back, looking angry.

I was supposed to pick up two bee nucs (starter hives of live bees) right about the time that Alex went into labor, but I messaged the bee person and told him why I was running late. I felt very conflicted as I really needed to pick up the bees and bring them home, but I was also worried about Alex. I checked her temperature, and it was only 101.5°F, which is a little on the low side but probably not low enough to assume she had milk fever. She wasn't eating until we gave her Power Punch, which is an electrolyte and molasses concoction. I couldn't check her for ketones because that would require catching her urine, and she wasn't standing up, so I'd have no idea when or if she peed.

She hadn't passed her placenta, and there was not much hanging out in terms of membranes, and she was still pushing every few minutes, so I was not completely convinced that she didn't have another kid inside. I've never seen a doe really "push" if the only

The kid was bopping Alex under her chin because he wanted to nurse. Newborns instinctively know that they can find milk above their head, so when a doe is standing, they will use their nose to tap the underside of the doe's body and udder. Most learn quickly where the teats are located and will go straight for them when they're hungry. Due to Alex's condition, however, she wouldn't stand up, so the kid couldn't find milk.

thing left was the placenta. I did an internal exam to check for a retained kid, but there wasn't one.

I eventually was able to milk out her colostrum and, because she wouldn't stand up and let the kids nurse, I gave it to the bucklings in a bottle. At least they were doing well.

I called the University of Illinois vet hospital and spoke to a vet who said that it sounded like Alex might just be in pain because the kids were so unusually large. So, I jumped in my car and went to pick up the bees and bring them home. There was a farm supply store on my route, so I stopped and picked up some CMPK to treat for possible milk fever, which is caused by a calcium deficiency.

The next morning I knew something was wrong. Her temperature was down to 99°F, which meant either that it was milk fever or her body was shutting down and she was dying. I put her in the car and headed for the university vet hospital.

Within a couple of hours, after an ultrasound and blood tests, we learned that she had metritis and was acidotic. Thankfully, she didn't have ketosis or hypocalcemia (milk fever). They thought that they might see a very small mummified fetus in her uterus, but they weren't sure because you can only make educated guesses on an ultrasound. They said she probably had the metritis before she went into labor, which would explain why she had such wimpy contractions and why I had to pull the kids and why she still hadn't passed her placenta twenty-four hours later. They still didn't have all the answers, though, and Alex was in terrible condition, so I left her at the vet hospital and headed home.

After I left, a couple of reproductive specialists did another ultrasound, and they didn't think there was a retained mummified kid. They thought it looked more like a uterine tear. But they were all in agreement that she had had an infection before she went into labor. The vet said she'd gotten worse since I left, so they were actually giving her morphine on top of the Banamine. They also started her on an oxytocin drip, hoping that would help her expel the placenta.

I think the reason her temp was so low that morning was because she was shutting down. She really looked like she was just ready to die, so it was hard for me to imagine that she could look worse. They said that if she had a uterine tear, the day after the birth was always the worst, and that if she was going to make it, she should improve a lot by the following day.

The next day they had a reproductive specialist do an ultrasound and a vaginal exam, and he said that she had an interior vaginal tear that went all the way through into her abdomen and that it could ultimately cause more problems than a uterine tear. The vet said, "Goat uteruses love to tear. It's just what they do. But they're also really good at healing themselves." Vaginal tears can wind up with a lot of scar tissue that would cause kidding problems in the future. If we wanted to breed her again, they suggested bringing her back in thirty days for another exam to see if scar tissue was building up and looking problematic. That was another first for me!

It was such a happy surprise that Alex not only survived but actually began to nurse her twin bucklings when they were a week old.

I'd never heard of an interior vaginal tear, but that's probably why the 5-pound, 2-ounce kid had so much blood on his head when he was being born.

Alex continued to improve, and I was able to bring her home after five days in the vet hospital. On the phone, the vet told me how much better she looked, but when I arrived, I thought she looked dreadful. After I brought her home and put her in the barn with her kids, she just lay there all the time with her ears back, and she'd still push every few minutes. The vet said that was because of the internal tear. Every time I looked at her, I thought that she might still die. I'd never seen a goat look so miserable, and I just kept reminding myself that we'd done everything possible.

I put her in the kidding stall with her bucklings, whom we'd been bottle-feeding. I hoped she'd tolerate them, but it seemed that she remembered them as she would lick them when they came near her. Three days after she came home, she finally started standing up to eat. As I was giving one of her boys a bottle one evening, the other one was bopping my leg, hoping to find a new spigot for milk. Since Alex had been tolerating them quite well, I decided to give it a try and see if she'd let the kid nurse.

I sat down in the straw and put the kid under her. I put the teat in his mouth while squeezing out a bit of milk onto his tongue. He immediately began to suckle! I supported his head for a minute or so, then I let go, and he continued to nurse, and Alex just stood there like a perfect mama! The other kid, who had already had a bottle, wasn't interested, so I decided to let him try nursing in the morning before giving him a bottle.

The next morning I was so excited about the possibility of the second kid starting to nurse but reminded myself that it might not work. *Don't get your hopes up too much*, I told myself. When I walked into the kidding pen, I checked the tummy on the kid who had nursed the night before, and it was quite full. I felt Alex's udder, and it was loose. We had been milking her, so I knew it would have had more milk if the kid had not been nursing.

I sat down and put the other buckling in front of Alex's udder and opened his mouth, just as I'd done for his brother. He also began suckling immediately! I was over the moon excited, and the song "Miracle's Happen" from *Princess Diaries* immediately began playing in my head. Indeed, if anyone had asked me if this were possible, I would have said that I never say never, but I wouldn't get my hopes up about kids starting to nurse at eight or nine days of age.

Alex has continued to improve, and we recently moved her and her bucklings into a nursery pen with two other does and their kids, and Alex immediately decided she should be the queen of this little herd. She took on the other two does in quite the head butting match and won.

We have no plans to ever breed her again, as I'm not going to trust a visual exam to determine whether or not she'd be able to give birth vaginally in the future. Her last lactation lasted three years, and she is 6 years old now, so we will just continue to milk her and see how long she can go.

Even though Alex was giving birth to a purebred Nigerian kid, her experience is one reason I warn people against breeding Nigerian does to bucks of larger breeds. I have heard of a few goats that were able to give birth to kids from breedings like that, but I've also heard of 7-pound kids and C-sections, and does that died due to kidding complications caused by excessively large kids. Even when breeding to a buck of the same breed, you can wind up with an unusually large kid. You don't want to purposely create a situation with such a high risk of failure.

We continued to milk Alex for three years, and then we stopped when we went on vacation. She was producing only 2 to 3 cups a day at that point, and we didn't feel like we could ask someone to come over to milk her every day for only 2 or 3 cups of milk. With such low production, she easily dried up when we stopped milking. She is still enjoying her retirement on the farm.

CAESAREAN SECTION BIRTHS

Prior to our first caesarean I'd only spoken to a friend who had a goat that had required a C-section. I had asked her how she knew she needed to take the goat to the vet. Her response was, "Oh, you'll know." But how? I pressed her for details, which she was not providing. Now that we've had two goats who've had C-sections, I can say that it was obvious in both cases. Because I always want people to be able to learn from our experiences, I wrote up very detailed posts about both of those situations.

Our first C-section

I knew it would happen someday. I only hoped to avoid that day for as long as possible.

Wednesday

9:00 p.m.: I notice a 4-inch string of mucus hanging out of Caboose's back end. Usually that means you'll see kids in a couple of hours, so I stay with her until past midnight. She isn't even acting uncomfortable, but I decide to sleep in the barn, because you just don't see that much mucus that long before you see babies.

Thursday

3:44 a.m.: Caboose wakes me up with a bleat that sounds serious, although it wasn't quite as long as most goat screams that mean the babies are coming. I pull on my insulated overalls, boots, coat, and hat, grab an armful of towels, heating pad, camera, and phone, and I go to the kidding pen and sit down in the straw next to her.

4:37 a.m.: I go into the barn office and lie down on the futon, hoping to catch a little more sleep because Caboose has done almost nothing for the past hour. Every fifteen or twenty minutes, she lets out a bleat that's about half as long as one that means she's really pushing. I spend the next hour feeling like a jack-in-the-box, as I pop up and look out the window from the office at Caboose when she lets out a short scream.

6:05 a.m.: The sun is coming up, and I still haven't had any sleep, so I decide to check on Caboose again. She seems fine, so I start doing chores, feeding all the other animals in the two barns. About every fifteen minutes, Caboose lets out a real scream that sounds like she is serious now. I run in, thinking I'll see a nose or hoof or something, but there is nothing.

8:00 a.m.: I decide to do a vaginal exam, thinking that the kid must be sideways or something. As soon as I feel a nose, I'm happy and assume that I'll be seeing a kid within about fifteen minutes, and I finish doing chores.

8:33 a.m.: Still nothing. I do another vaginal exam and realize that just beyond that nose is Caboose's very tight cervix. She has not been dilating. I immediately think of my friend's Nubian doe who had to have a C-section a few years ago because she did not dilate. I call the University of Illinois vet clinic and am quickly connected to a

The buckling needed a little oxygen after he was born, but he survived and grew into a beautiful buck.

professor. After hearing the history, he is concerned that she is not dilated and explains how to manually dilate the cervix. If that doesn't work, he says we're probably looking at a C-section. I try to manually dilate the cervix, carefully following his instructions, and don't make any progress.

9:42 a.m.: I go into the house to call another breeder, to see if she has any tricks up her sleeve before heading to the vet clinic. She does, and adds that it could take thirty or forty minutes, so I try again.

10:30 a.m.: I go inside and tell Jonathan to help me get a crate in my car to take Caboose to U of I for a C-section. Still no progress on dilating the cervix, and when I pulled my hand out the last time, there was hair covering my fingers, so I knew the presenting kid was dead—had been dead for quite some time and was probably responsible for the current situation.

11:05 a.m.: I'm on the road to U of I. Caboose is absolutely quiet. I wonder if she's died. I second-guess every decision I've made in the last eight hours. After half an hour on the road, I hear her kick, and my sleep-deprived brain says *She's not even in labor!* because she is no longer bleating.

12:45 p.m.: I arrive at the vet clinic. The crate is placed on a cart that is wheeled

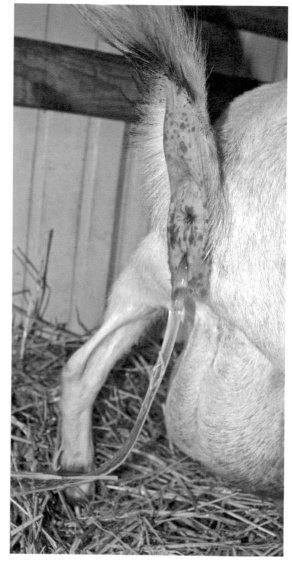

Goat owners often ask how much mucus they should expect to see before they know their goat is in labor. I don't get excited if I just see a little spot of mucus on the vulva, as it may mean nothing, but when you see this much mucus, you know the doe is getting really close to kidding.

straight to the operating room. I feel my throat getting tight and tell myself not to cry. She is going to be fine. Margaret arrives ten minutes later. She is in her senior year as an engineering student at the university. I called her when I knew she was done with classes for the day, and I told her what was happening. Caboose was her goat before she went off to college.

Everything starts moving really fast as more than a dozen people are buzzing around Caboose. A vet does a vaginal check and confirms what I'd said—she isn't dilated. It is obvious a C-section is the only answer for a positive outcome. Caboose is weighed and her belly shaved as the surgeon keeps reminding everyone that they need to hurry in case there is still a live kid inside. They shave her neck to insert an IV line. We discuss options for anesthesia, and one vet explains that gas will be the quickest, least stressful option for Caboose, so I agree. They put a mask over her face, and as soon as she is asleep, they put a tube down her throat. They shave her ears and attached tubes and wires. Six people surround Caboose closely: the surgeon and his assistant, the anesthesiologist and her assistant, and two students who are holding Caboose so she doesn't roll off the operating table. Half a dozen more masked veterinary students stand around the room to observe.

1:37 p.m.: The C-section begins. Within seven minutes the first kid is delivered and handed to a student who is waiting with a towel. I don't see any movement and know that no one knows if it is alive or dead. I watch as several people start working on the kid and overhear one say that it is alive. I wonder if it is a buck or a doe, but I know that it really doesn't matter. The kid's airway is suctioned, and everything is happening so fast, I can't keep track. I hear excitement back at the operating table and see another kid in a towel. All of those students who were just standing around are now busily working on our kids—drying them, administering oxygen, giving dextrose, injecting something else, suctioning the kids' airways.

I see the third kid get pulled out, and it is obvious this is the dead one that had been stuck in the cervix. The amniotic sac looks like it is filled with mud. As I suspected, the kid has been dead for some time. A student wraps it in a towel and sets it aside. I look at it and see that it was a buck. I want to know exactly what happened and why, but as I start to look at him more closely, I realize a fourth kid has been delivered. At some point, I hear the anesthesiologist say, "Get me a crash cart" and know that's not good. I leave the dead buckling and walk over to the kids that everyone is trying desperately to save. The anesthesiologist looks up at me and asks if it's okay to intubate the kid that they're losing. I nod and say, "Yes, that's fine." A few minutes later, she asks if I want them to keep trying. I ask if there's a heartbeat, and when she says no, I respond, "That's okay. We have two kids."

For the next two hours, the kids are the stars of the hospital. The little doe finds her voice, which brings everyone within earshot. People squeal like children when they see the kids—the little doe learning to walk and the little buck wrapped in a "bear hugger" to bring up his temperature, which has dropped to 91.6°F (10 degrees below normal). No fewer than five people say they want to take the kids home with them. Professors, students, receptionists, and custodians are all captivated by these tiny little angels.

5 p.m.: Margaret and I finally decide to leave the vet hospital. Caboose is awake and now has a walking epidural, so she won't be in pain, although she attempted to stand once and was completely unsuccessful. The little doe is a champion nurser. The little buck never quite figured it out, so they fed him some of Caboose's colostrum through a stomach tube. It was hard to leave them at the hospital, but the surgeon assured us that they would be carefully monitored at least hourly.

If Caboose can do all the normal goat things like walk around, eat, and drink, she and the babies can come home Friday.

I kicked myself quite a bit after this experience. Had I taken Caboose to the vet hospital sooner, we might have been able to save the third kid. On one hand, I told myself that I should not have second-guessed the vet's advice and called the other breeder and continued to try to dilate the cervix manually. On the other hand, I told myself that I live two hours away from the vet hospital and if I could have delivered the kids at home two hours earlier, that might have saved one. But when you are in the middle of a birth, you have no crystal ball telling you which decision will give you the best outcome. It is a classic example of how hindsight is 20/20, as the cliché goes. Arguing with yourself about whether you should have done this or that does no one any good. No one knew that the fourth kid had been dead for days and was blocking the exit for everyone else. I've learned that kids who die in utero days or weeks before birth have a way of complicating the birth, and it's very frustrating.

Although many breeders talk about always being successful at dilating the cervix manually, the textbooks say that it is rarely successful. The textbooks also state that it is challenging to figure out how often manual dilation is successful because you will never know when the doe would have actually finished dilating without assistance and when it was actually needed. Since I have had no other instances of failure to dilate, I suspect many owners tend to attempt this much sooner than would researchers, who are not emotionally attached to the goats. In other words, the owners who are so successful didn't have goats that were really having a problem dilating. The goats were just progressing more slowly than the owner expected them to.

If a doe has been in labor for too many hours, the placenta will start to peel away from the uterus, and depending upon where a kid's umbilical cord is attached, they will either be fine or they will die. The vastly different condition of each of Caboose's kids following her caesarean illustrates how that works. The one kid had been dead for days, as shown by the fact that its hair was falling out. One kid was born in distress and died within a few minutes of birth. One kid was born in distress but came around within a couple of hours with a lot of medical assistance, and one kid was born perfectly healthy and screaming within a few minutes.

According to the vet, Caboose's placenta had started to separate because she had been in labor for many hours. The umbilical cord of the kid that died within a few minutes of birth had been attached near the place where the placenta was separating, so it was oxygen deprived and distressed when it was born. As the vet told me later, it's probably better that the kid died because it might have been blind or had any number of other problems. The other two kids had umbilical cords that were attached to the placenta in a place where it was still attached to the uterus and allowing for blood flow (and oxygen) to the kids.

Like most goats who have a C-section, Caboose's incision was on her left side, starting a few inches from her spine and going straight down several inches. The placenta was left attached to her uterus and allowed to gradually expel on its own over the next few days, which surprised me, as I had thought they would remove it during surgery. But I've since learned that ruminant placentas are almost never manually removed, even if they take days to expel. She was given a long-acting antibiotic to prevent infection of the incision, which would also prevent infection caused by the placenta's delayed exit. Caboose's recovery went well. When we brought her home, we put her and her two kids in their own private stall, and within a couple of weeks, we decided to reunite them with the rest of the herd.

Caboose fully recovered and got pregnant and gave birth to a huge single doeling the following year. After that, however, she never got pregnant again, even though she came into heat multiple times and was bred by a buck that was settling other does. She was officially retired at age 9, which was a year after she gave birth to the single doeling. Both of the kids born by C-section grew up to be healthy, productive animals.

Lesson learned via C-section

B·L·O·G
THU
FEB 20
2014

A week ago Monday it was obvious that Giselle was in labor. She woke us up over the video monitor shortly after 6:00 a.m. but wasn't really making much of a fuss until closer to 8:00, which was when I finally went outside to the barn because I thought I saw mucus glistening under her tail. When I got to the barn, I didn't see any mucus, so either it had already fallen into the straw or there was some sort of optical illusion over the video monitor.

I spent most of my time sitting in the barn office watching Giselle through the window because I didn't want to go outside into the zero-degree temperature until I really needed to be out there. Around 9:00, Giselle seemed to be seriously, but quietly, pushing, and from the office I could see a kid's hoof presenting.

I left the office and went into the barn, assuming that the kid would be born fairly quickly. However, once I got within a few feet of Giselle, I knew I had trouble on my hands. The hoof was much too big. I went back into the barn office, called Jonathan's cell phone, and asked him to bring my insulated overalls. This would not be a quick birth, and I was already freezing. He brought out the

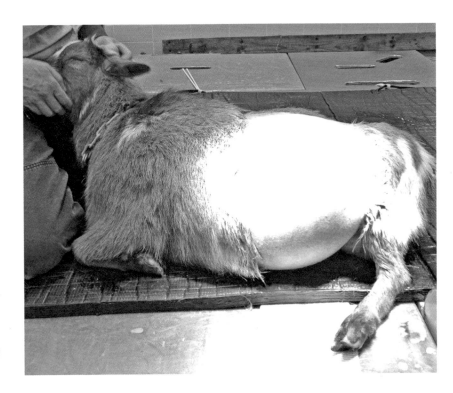

Giselle's left side was shaved, and she was given an epidural in preparation for the surgery. Since we had to use the cow-sized table, the vet student holding her was also sitting on the table with her.

overalls, and I asked him to stay with me because I would probably need his help. At this point, the hoof had been sitting right there, sticking out of Giselle a couple of inches for at least twenty minutes and hadn't progressed at all. I grabbed several vinyl gloves and a bottle of iodine, put on my insulated overalls, and headed back into the barn.

Jonathan held Giselle while I put my hand inside her trying to figure out how to get the kid out. My hand could not slide between the kid's head and Giselle's pelvis. As I slid my hand in, the head moved back into her uterus. I looked for the second front leg, but there was no room for me to maneuver inside her, and I was having no luck. Normally, I don't worry about pulling a kid when there is only one leg presenting, but it was obvious that this was a very large kid, and Giselle is a smaller than average doe. I didn't want to deliver the head and one leg, only to have the kid get stuck on the shoulders. With a kid that big, I wanted the assurance of having both legs front and center. I tried a second time. I did grab what I thought was another foot at one point, but I didn't think it was a front leg. *Third time is the charm,* I thought as I tried again. The kid is too big. Her pelvis is too small. I told Jonathan that I was taking her to the university vet clinic. I ran into the house to phone and tell them I was coming while he

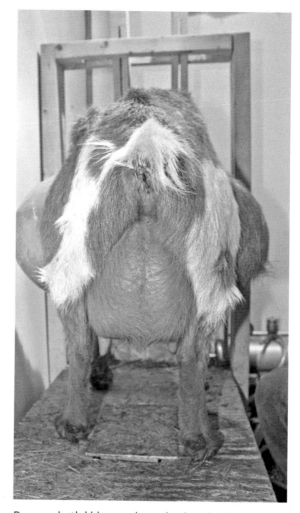

Because both kids were born dead, we had to begin milking Giselle immediately after birth. Within a few days after surgery, she was able to jump up on the milk stand as usual. You might notice that her left side has no hair after being shaved for the surgery. She also continued to look pregnant for the rest of her life because she had lost her girlish figure a few years earlier. Whenever people would see her in the pasture, they would always ask when she was due to kid.

got my car ready. It's a small SUV, and we put several empty paper feed bags in the back with a blanket over them. Jonathan carried Giselle to the car, kid's leg still hanging out of her back end. In her condition, I knew she would not jump over the back seat to join me.

For the entire two-hour drive to Urbana, every time Giselle let out a scream, I kicked myself for one thing after another. *I should have just pulled harder. I don't need to be taking her to the vet clinic. I should have had more confidence in myself.* And then there was the "should" that I repeated the most. *I never should have bred her again.* I have a two-strikes rule: if a goat has a kidding problem twice, she's retired. Unfortunately, I had rationalized and talked myself out of retiring Giselle the previous year when she had a hard time delivering a fairly large kid.

They were extremely busy at the university vet clinic, and there was not the usual large group of students meeting us. In fact, it was one senior vet student who met us, and she was in the midst of her first day of clinic rotations. Luckily, I'd been to the clinic enough to tell her how we needed to move Giselle inside (using a cart) and how to open the sliding door, and so on. The resident joined us fairly soon, and I explained to her what had already happened. She examined Giselle and asked how I felt about a C-section. I told her that I was expecting it. "This is what I get for breaking my two-strikes rule," I told her.

The mid-size operating room was being used for surgery on a small pig. Giselle was clearly too wide for the smaller operating room, so they wheeled her into the large operating room, which has a cow-sized table. Definitely overkill, but it worked. The senior vet student assigned to Giselle had never met a goat before and was full of questions. Giselle was an excellent ambassador for the breed, being very cuddly and agreeable, in spite of her condition, because her labor had essentially stopped.

After finding people, gathering supplies, getting drugs, shaving Giselle, and administering an epidural, it was more than two hours before the first kid was delivered, and it was dead. I was watching from an open doorway and thought I'd heard a squeak, but when no one said anything about the baby, I finally asked how it was doing. One of the students looked up at me and shook her head.

"It's dead?" I asked. "Yes, but we're working on the other one now." That was the kid that was engaged in the pelvis, and I heard the vet professor tell someone to push the head and foot out of the pelvis and back into the uterus. A minute or two later it was also delivered dead. The professor said it was tough to get the kid out of the pelvis because the head was really jammed in there. As we had all suspected, there was no way it could have been born naturally. It was simply too big.

Everyone had been in awe of the large hoof sticking out earlier. Now everyone was talking about what a huge kid had been delivered. The kid blocking the exit was a 5-pound buckling. Most Nigerians are in the 3-pound range. Although we have had a 5-pound kid born once before, it was to a doe that was 22 ½ inches tall, and Giselle was 3 inches shorter than that! A 5-pound kid was simply too big for her to deliver. And the poor little 3½-pound doeling never had a chance with her big brother blocking the exit.

Once the kids were delivered and the vets assured me there were no more, I decided to go get lunch. It was close to 4:00 in the afternoon by then, and I hadn't eaten since 7:00 that morning. After eating at the local health food co-op, I went back to the university to see how Giselle was doing and was amazed that she had won so many fans already. "She acts just like a princess," said the senior vet student, as she described how Giselle seemed content to sit and let everyone take care of her.

Seeing Giselle's side completely hairless after being shaved for surgery, I couldn't imagine putting her back into a sub-zero barn, so I stopped at Lowe's in Bloomington to buy a vinyl-flooring remnant to put in the barn office. Giselle could stay in there for a few days until the weather hopefully reached saner temperatures.

And on the drive home alone I kept repeating to myself that my two strikes rule needed to be followed in the future. There would be no excuses for any goat, regardless how much I adore her and want more babies from her. It simply is not worth it.

The take-away from this story is that a goat who has had kidding problems twice should be retired. Giselle had a hard time as a first freshener, but I felt that it was because she had been too small when she was bred. She was only 35 pounds after the kid was finally born, and that's too small. Because of her, I don't breed does until they weigh 40 pounds.

You already read about Giselle trying to give birth to the dead kid that was presenting ribs first. The following year she was actually able to give birth to a 4½-pound buckling with no trouble, but the year after that, she again needed assistance. Because she has a buttery soft udder and the longest teats in the herd, I really wanted another daughter from her with the same wonderful mammary system. And I kept rationalizing and telling myself that she just had bad luck. I had always said that I had a strict two-strikes rule, but the truth was that I'd only ever had one goat with two strikes before, and I wasn't especially fond of her, so it was easy to retire her. The two-strikes rule is now firmly in place and will not be violated ever again on this farm.

5

CHALLENGES WITH NEWBORNS

Although many new goat owners worry about the doe, few mention being concerned about the newborn kids. Concerns related to kids usually seem to center on the doe. What do you do if the doe rejects the kids? What if the doe doesn't have enough milk? What if the doe dies? The answer to all of these questions is that you would have to bottle-feed the kid unless you were lucky enough to have another doe on the farm that would accept them, which is unlikely.

Just as things can go wrong with the birthing process, kids can present a whole different set of challenges. This section is devoted to stories about kids that were either born with problems or had problems caused by the birthing process.

Sucking disorder in a goat kid?

B·L·O·G
THU
JUN 12
2008

It makes sense that if something can happen in one species of mammal, it could happen in another species of mammal, right? One thing that most people don't know about me is that I was an International Board Certified Lactation Consultant and a La Leche League leader in a former life that ended in the late '90s. One of my mantras was "Breastfeeding is not supposed to hurt—if it hurts, something is wrong." Often, the baby did not have enough of the nipple (and breast) in his or her mouth, which made the end of the nipple red and sore. The same thing is now happening with one of Star's kids, and Star is not liking it. I don't blame her.

Normally, she could be nominated (and probably win) Goat Mother of the Year. She does things that I don't see other does do. For example, after a kid has nursed for a while, she walks over to another kid and encourages it to nurse. Most goat dams don't pay attention to who is nursing, as long as it is one of their kids. Star is an outstanding mother, but this is trying her patience. I noticed yesterday that the tiniest of her doelings was not nursing correctly. Star has very long teats, about 2 to 3 inches long at rest. This is pretty long for a Nigerian dwarf. Her kid is tiny — less than 2 pounds. The little doe is sucking on only the tip of the teat. Yesterday, I looked at the tip and noticed that it was turning red, which is clearly a sign that something is wrong because this goat has black teats. I also could not hear the doeling swallowing, which is another sign that she is not nursing correctly. Today, the little girl still is not nursing correctly, and Star is starting to side-step when the little doe latches on, which means she is pulling the teat out of the doeling's mouth. In other words, she doesn't want to let the doeling nurse. She is still completely willing to let the other kids nurse, though.

When the little doe nursed, she was sucking on only the tip of the teat. There is quite a bit of space between her nose and the udder. When a kid is nursing correctly, the tip of the nose is up against the udder. When a human baby nurses correctly, the tip of the nose also touches the breast.

I used to see the same problem in human babies, usually after they'd had a bottle or two or three. Bottle nipples for human babies are much shorter than a human nipple when it is inside a baby's mouth, so when human babies suck on an artificial nipple, some of them get the idea that they should be sucking on something at the front of their mouth. In the case of this little girl, I think her diminutive size is the main culprit. Her mouth is just too tiny for her mama's long teats. She'd probably be perfectly fine if she were the kid of a first freshener that had tiny teats.

But Star is her mommy, and we have to play the cards we're dealt. Contrary to what I would have done with a human baby, I have given the doeling a bottle three times now — once last night and twice today. The Pritchard teat is a long, soft nipple that is pretty similar to a goat's teat. In this case, I think it might be helpful because this little doe obviously doesn't like the idea of something farther back in her mouth. The Pritchard teat is long and hopefully will help her get used to the idea of sucking on something longer. Tonight I heard her swallow for the first time when she was nursing on her mama, so I am optimistic that she'll make more progress overnight.

The little doeling did learn to nurse correctly. Within a day or two, she was getting all of her nourishment from her mother. Prior to this experience, I had never understood why a doe might be unwilling to let one of her kids nurse while letting the others do so. If a doe is unwilling to let any of

It is not always easy to see, but this photo shows a kid nursing correctly, with his tongue covering his lower teeth.

her kids nurse, there is probably a hormonal problem (which can happen with premature births), or it is simply a behavioral issue with the dam. However, if she is refusing to let only one of the kids nurse, there could be a problem with the kid. In addition to situations like this one, a kid might also have a cleft palate. Just like in human mouths, the roof of a goat's mouth should be solid. If it is not, the kid could have trouble nursing, and you might also see milk coming out of its nose when it does nurse.

Recently I've seen people suggesting that the teeth of some kids are too sharp when they're born, and they talk about using a file on them. This is not necessary because when a kid is nursing correctly, the tongue is covering the lower teeth, so they are not coming in contact with the doe's teat. If their teeth were too sharp, they'd be hurting their own tongue. There is no way that you can make the teeth "not sharp," and luckily, you don't need to do that.

While most tiny goats eventually catch up to their siblings in size by a couple of months or a couple of years of age, Lil Dipper remained tiny forever. She still lives on our farm, and at 11 years, she is still the height of most 6-month-old goats, although she's very meaty. I didn't sell her partly because I didn't want her to ever be bred (due to her diminutive size) and partly so that she could keep her mother company in her retirement years. Star was my first milk goat and had a very old and obscure pedigree. Star died in her sleep at age 14½. Lil remains with us, and I point to her as reason to buy from known pedigrees.

Cheating death

I came home last night to learn that Beauty had given birth unexpectedly to triplets and the little doeling died. She was small and soaking wet when Katherine found her. Only the mouth and nose had been cleaned off by the mother. No doubt she died from hypothermia. When people ask me why we have to be at the births, I tell them that 99 percent of the time we just need to make sure the kids are dried off. It is amazing that something as simple as a towel can be a lifesaver, but it is.

This morning I went out to check on the two boys. The white one was lying alone, and the buckskin was under the heat lamp. I felt their tummies, and they were not full. I gave the mom some alfalfa and stood the kids up under her to nurse. The mom started doing a tap dance, moving from side to side as she munched the hay. The white buckling could hardly stand and gave up quickly. The buckskin followed his mom from side to side but then lay down under the heat lamp to rest. After about fifteen minutes, she finally let him have a few sucks. I am not in the mood to have any goats die, so I decided to take the boys inside and bottle-feed them.

That last comment sounds like Twilight Zone fodder, but as I was walking into the house with the kids under my coat, I remembered my experience with Carmen. I had never had a kid so near death, so I called people with more experience. Some had the attitude that if the kid couldn't make it on her own, it wasn't worth saving her. I couldn't just let her die, and this morning, I couldn't let these boys die.

The white kid was already suffering from hypothermia. When I opened his mouth to put the bottle nipple in, I realized his mouth was ice cold. No wonder he had no energy. He was almost dead. I was ecstatic that he could still suck. (Carmen couldn't when I found her, so I had to tube feed her.) Not only could this little guy suck, but he made it quite clear that he was starving! I had warmed up 4 ounces for the boys, which should have been more than enough for two twelve-hour-old kids. The little white one sucked down most of it without stopping. I offered the rest to the buckskin, deciding to give him more after I'd tried to warm up the white one.

As I had done with Carmen and other kids suffering from hypothermia, I put him in warm water—and I got a surprise. When his lips touched the water, he started drinking it! Although he had no energy for anything else, he knew he needed nourishment and fluids. That's a great sign. I rubbed him in the warm water, and he moved his legs. After about five minutes, his mouth felt a bit warmer, so I lifted him out of the water and wrapped him in a towel with only his nose sticking out. I got the heating pad and plugged it in. I placed him on it and covered him with the towel. I repeatedly checked his temperature because I didn't want him to get overheated. When his mouth was only cool, I uncovered him. I'm hoping that between feeding and lying on the heating pad, he will warm up and totally come back to life.

I hate to ever say anything about whether or not I think a kid will make it. I've lost some, and I've saved some that were in worse shape than this little boy. I am guardedly optimistic, hoping that when I get home tonight, I'll be met by a pair of bouncing baby goats in my living room.

You save some

And you lose some. We were quite hopeful about the little bucks that we brought into the house yesterday. They were eating well, and the little white one had even started walking. But sometime between the 3:00 a.m. feeding and sunrise, he died. The buckskin is still alive, but he is not very strong. Neither of these kids weighed even 2 pounds, and just like human babies, low birth weight is a tough handicap to overcome. But at least we know we've tried. I'm glad we brought them into the house. When milking their mother, we get very little milk, so even if she had been willing to let the boys nurse, they wouldn't have been getting enough. The tiny udder, low production, and lack of mothering instincts makes me think something is wrong with mom hormonally, but what?

This birth is a perfect example of why we try to be with every doe when she gives birth. Although this doe was not a very good mother, it is usually impossible for even the best mother to be able to clean up three kids before one suffocates or gets hypothermia. Although I was confident that the doeling had died from hypothermia when I wrote the original post, it is also possible that she was born too quickly after one of her brothers and the mother was simply not able to get her nose cleaned off before she suffocated. It makes more sense that the dam cleaned off her nose and then realized she wasn't alive, so she stopped cleaning her at that point.

Unfortunately, I can no longer find my breeding records from this year, so I don't know if this doe had been pen bred or if we had a due date, but seeing three kids under 2 pounds each makes me think they were premature, which would also explain the mother's lack of maternal instincts and almost non-existent milk supply.

Today I know that if a kid has hypothermia, it should be warmed up first and then given milk. If the body temperature is below 100°F, the digestive system isn't functioning because the body is putting all of its energy towards more important things like the brain and lungs.

The buckskin buckling survived, and I castrated him and sold him and his mother to a pet home. Because of the birthing issues experienced by the doe—and the fact that she had given birth to dead kids a month premature a year earlier—I decided she should not be bred again. Today I know that our herd was suffering from severe copper deficiency back then, and premature birth is one of the symptoms.

A blind kid?

B·L·O·G
MON
MAY 3
2010

The past few days have been crazy. We've been gardening, scything fresh grass for the mama goats in the barn, and dealing with three goats who freshened. Although we did not spend hours with the mama goats in labor because all three of them surprised us, we have been spending a fair amount of time with one of the kids.

He was born on Friday to Annie Oakley, along with a very big brother and sister. He was only 1½ pounds at birth, but let's take a couple of steps back in time. Jonathan had left a gate open, and the horses were out in the road. Katherine had just put them back where they belong, and as she was walking past the pregnant doe pasture, she saw Annie walking along with a kid's body swinging behind her. Yes, the kid was swinging by its neck from her back end.

The kid's head was still inside. Katherine climbed over the wood section of the fence, screaming for someone to come help as she ran to Annie, grabbed the kid and pulled it out, which was quite easy since it was so small. There was a huge kid stumbling around the pasture already, so we're thinking that Annie didn't realize this little sprite of a kid was even being born. After giving birth to a 4-pound kid, she probably wasn't pushing much to birth the 1½-pound kid, and he didn't weigh enough for gravity to help, even though Annie was walking around.

When she pulled him out, Katherine thought he was dead because he wasn't responsive, but after she rubbed his body a little, he started sneezing and moving. Twenty minutes later, a third kid was born, a doe that weighed 2½ pounds. Katherine and Mike moved them into the barn, where the big buck and the doe did what kids do—they stood on wobbly legs and started staggering towards their mama and nursing. The little kid never stood. After an hour, Katherine decided to bring him into the house for a bottle.

He didn't stand for 24 hours, so bringing him inside was the right call. If a kid can't stand, they can't nurse, so he would have starved. He wasn't doing a very good job holding his head up, and when I called him NoodleNeck, Katherine decided that Linguine would be a good name for him. He didn't attempt to walk until yesterday, which was when he was two days old, but things got really weird this morning. The little guy has been sleeping in a box in Katherine's room. This morning, she put him on the bed with her, and he tried to walk right off the edge of the bed several times. It was as if he didn't see where he was going.

"Mom, I think this kid is blind," Katherine said this morning.

Throughout the day, we've been coming up with test after test to see if Linguine can see. If we put him on the floor, he walks into furniture and walls. If we put him on a couch or bed, he walks right off the edge as if he doesn't know it's there. (Of course, we catch him.) We flick our fingers at his eyes, and he doesn't blink. When Katherine puts the bottle in front of his face, he does nothing. As soon as she touches his lips, he tries to grab the nipple.

Of course, my first assumption was that we were simply being paranoid. Why would he be blind? He looked perfectly normal and was acting healthy. Then I remembered his birth. How long had he been hanging from Annie's back end? I asked Katherine if the umbilical cord was already broken when his head was trapped inside his mother. Yes, the cord had been severed, which meant he wasn't getting any oxygen until she pulled him out. I googled "oxygen deprivation and blindness." Yes, oxygen deprivation at birth can cause blindness, deafness, and all sorts of mental and cognitive problems in children, so why not goats?

Now, the question is, what do you do with a blind goat?

Most people think that the only two outcomes are live or dead kids. They think that if they hurry, they'll be guaranteed a live kid. They don't realize that if a kid has a challenge like this at birth, it could wind up blind. After Caboose's C-section, when one of the kids died shortly after birth, one of

This is a photo of some kids that were brought into the house for bottle-feeding because their mother was not doing well after kidding. If you think it looks like the kids could jump out of here, you're right. They can. The laundry basket is usually being used for kids that are not doing well. Not only do they not think about jumping out, but they're not strong enough to do that. In Linguine's case, he couldn't see the world outside the laundry basket, so he stayed in there longer than most. In general, once kids can jump out, it means they're strong enough to be moved to the barn.

the vets said to me later that it was probably not a bad thing that it died, because if it was that close to death when it was born, it could have been mentally deficient. It would probably have such serious challenges that we might have had to face the possibility of putting it down later. This is one of the experiences that actually help me to stay calm during a birth now. I don't need to hurry, because a kid doesn't just go from perfectly healthy to dead within seconds. The placenta gradually starts to peel away, and the kid's body starts to shut down system by system.

Linguine update—the not-so-blind goat

B·L·O·G
FRI
SEP 17
2010

Back in early May, a kid was born who appeared to be blind. We had to bottle-feed him because lack of vision was not his only problem. With time, he grew stronger, and he appeared to have some vision. When we first moved him from his laundry basket in the house to a pen in the barn, he would lie down with no regard to where he was facing. Generally, a goat always has its back to the wall so it can see what's going on, but not Linguine. We'd often see him lying in the straw facing the wall.

It took him almost a month to start grabbing the bottle, even if we rubbed the nipple against his lips. When he was 2 months old, we'd take him out of his pen and go walking with him. He would follow along fine, as long as we were not more than a couple feet away from him. If he got interested in something and stopped to smell the roses and we kept walking, he'd be screaming. When Katherine took him walking, she would loudly shuffle her feet in the gravel driveway, which seemed to help him keep up with her. We finally figured that he was incredibly nearsighted and couldn't see beyond a few feet. But that was better than being completely blind. Still, what could I do with a goat that was less than perfect? I didn't feel that I could sell him, and I didn't want to give him to some stranger, who might then eat him or take him to a sale barn to make a few bucks.

In July, one of my blog readers purchased a buckling who

needed a wethered companion, and when her children met Linguine, it was true love. Well, who wouldn't love a little goat that wants to stick to you like glue? Knowing the family, I felt comfortable giving him to them. And seeing how much the children loved him, I knew he'd have a good home.

I recently received an update from Linguine's new owner. She said that he's doing well, and she thinks his vision is continuing to improve, which is really great news!

Lessons in neonatal goat care

Last night was the final night for our first winter intern, so we decided to take her out to dinner. Of course, as soon as I told her we were going out to dinner, I went into the kidding barn to discover that Sadie was in early labor! We decided to run out right away so that we'd be sure to get home in time for her kidding, especially because she was a first freshener.

When we arrived home from the restaurant, Sadie was still in early labor, so I checked email and did some paperwork and decided to get to bed by 10:00, which is early for me. We now have a video monitor to keep an eye on does in the kidding pens, so I was watching on the TV in my bedroom. I was in bed reading an article about dealing with drought when I heard a soft sound that was unmistakably the sound of a doe pushing. I jumped out of bed, ran across the hall to tell the apprentice that Sadie was pushing, and I went back into my bedroom to put my clothes back on. As I was in the middle of getting dressed, Sadie changed her tone and was in full-blown screaming panic! My son could hear her in the basement, which is two floors below the TV, and he said it sounded like she was being ripped apart by coyotes. The apprentice and I went into slight panic mode and went running outside with only one layer on our lower halves, which we didn't even realize until we were in the barn and our legs started to feel really cold.

Within five minutes of our arrival, a black kid was presenting. Just as the apprentice said, "Oh, this one's breech," the kid shot out. I wiped off the nose, and realized Sadie was ignoring the kid. I held it in front of her face, telling her, "Look at your baby, Sadie. You need to lick your baby." She was ignoring it. I didn't want to wipe too much because I wanted to encourage Sadie to bond with her. However, my guess of the kid's size was about 2 pounds, which is quite small and very vulnerable to hypothermia, so getting the kid wiped or licked right away was important. After a few minutes, Sadie started to lick it, and then she started to push again. After only a couple of pushes, another small kid came shooting out, this one nose and hoof first.

After a few days we discovered that the brown doeling in the back was blind and had a number of other problems.

By now Sadie was licking the first kid, and I was starting to dry off the second one, and Sadie let out a little noise. I saw a third kid presenting and handed the second one to the apprentice to dry off as I caught the third one, which also came out very quickly, also breech. A few seconds later, Sadie let out another yell, and a big bubble came out. It was dark, and both of us thought we saw kid parts in the bubble, but it was just fluid, which was good because we were running low on towels! Sadie had not been very large, and we were not expecting more than two kids.

Although the doelings were all quite small, they were very healthy and trying to walk around the pen within minutes. The second one, which was brown, was especially active, and I was calling her "our little world traveler." I realized that Lizzie was paying very close attention to the kid through the pig panel that separates the kidding pens, and it occurred to me that Lizzie is an outstanding producer, while Sadie is a yearling and will probably not be able to produce enough milk to feed three kids. Lizzie also happens to be Sadie's mom, so it seemed totally appropriate that she would nurse one of her granddaughters.

"I've never done this before," I said, "but I think this could work. Lizzie is so interested in this kid, and she just gave birth earlier today. I bet she'll accept this one as hers." So, I took the little brown kid into her pen and let her sniff it, saying, "Lizzie, you wanted a daughter, didn't you?" Then I placed the kid under Lizzie and tried to get her to nurse, which was fine with Lizzie, but the kid was oblivious. She had zero sucking reflex, and in spite of my efforts and Lizzie's patience, I realized this was not going to work simply because the kid wasn't interested in nursing.

But kids need 10 percent of their body weight in colostrum within twenty-four hours, and they need half of that within the first few hours after birth. I had called over the TV monitor for Mike to bring my camera, and when he arrived, I asked him to bring me a milk bucket, bottle, strainer, and funnel so that I could milk Lizzie and give colostrum to the kids with a bottle. At this

I made these kid coats from a sweatshirt sleeve. The wristband became the kid's collar, and I cut two holes in the bottom for the front legs. You should use kid coats for only the first day or two with kids that are having trouble maintaining their body temperature. It is always possible that the coat could get caught on something and cause injury or death of a kid.

point, the kids were an hour old, and none of them had any sucking reflex at all.

I was on my knees milking Lizzie as she stood in the middle of her kidding pen until she lost patience with me, but I had 3 ounces, which would be a good start if I could get that into the kids. After trying each kid multiple times, I finally got an ounce into the black one, and a little later, I got another half an ounce into her. Sadie let me milk her, and I added another ounce of colostrum to the bottle. After getting an ounce into the brown one, but nothing into the cream one, I said to the apprentice that I was taking the cream one into the house because I needed to get some colostrum out of the freezer, and besides that, I was freezing.

When we got into the house, we realized it was 1:30. We had been outside for three hours, and Sadie had given birth within the first ten minutes after we got out there! I was thinking out loud about why we had three apparently healthy kids that were not sucking, and everything finally started to add up. I put my finger in the kid's mouth and realized it was not warm. It wasn't ice cold like a kid with severe hypothermia, but her body temp was definitely below normal. When kids get chilled, sucking is one of the first things to go.

By the time we had warmed up the colostrum for her, she was already warmed up quite a bit from simply being in the house, and she sucked down the ounce of colostrum like a pro. My theory now verified, I went searching for kid coats. I found only one, so I chose a sweatshirt to sacrifice and made two more, then took them outside to put on the other two kids. I wasn't terribly worried about them because they had already consumed a decent amount of colostrum, but I wanted to make sure they stayed warm enough. I went back inside, put the cream kid into a laundry basket, and went to bed.

I knew the other kids would need more colostrum before morning, however, so I left the TV monitor on. At 4:30 I was awakened by one of the kids and Sadie talking to each other. I woke up Mike and updated him on everything that had happened, telling him that he needed to get more colostrum into the kids. He went outside and had no luck getting those two kids to nurse, and just as he came back inside, the kid in the laundry room was waking up. I wanted to help but still hadn't really warmed up myself. Mike defrosted another ounce of colostrum for the baby in the laundry room and she sucked it down. He handed her to me, and she started bopping my chin and trying to suck on it. I got excited and suggested he take her outside and see if she'd nurse on her mom.

Mike went back outside with the kid, returning about an hour later at 6 a.m. to tell me that the cream kid never nursed on Mama Sadie, but the other two did finally! Because the cream kid had just consumed another ounce of colostrum from the bottle I wasn't too worried about her. All morning, we've continued to check on them, trying to make sure they're nursing or taking some more colostrum from the bottle. Mike got another 3 ounces from Lizzie. Once two kids are successfully nursing on Mom, we'll bottle-feed the third one. The black one appears to be very good at nursing, and the cream one is very good at taking the bottle. Not sure about the brown one yet.

We said good-bye to our apprentice this morning, so now it's up to us to get these little does started on the right track. I'm afraid we still have a bit of a challenge ahead of us.

Although the black kid and the cream kid learned to nurse and grew up to be big, healthy kids, the brown doeling continued to have problems. After a week or so it became obvious that she was blind. When we tried to leave her outside with her mother and sisters, she kept winding up in the water bucket. Although it was a short 2-gallon bucket that she could easily get out of, it meant she was soaking wet when it was still quite cold outside. She also was not as "smart" as the other kids, who would lie under the heat lamp.

Initially she would fight the bottle like no kid I'd ever seen. For weeks we had to strong-arm her and hold the nipple in her mouth, just like we have to do for a newborn who doesn't understand how to take a bottle. She got a little better but then would get worse again. It was obvious that she was not "normal," but I wasn't sure what to do with her. It seemed that she had some type of sensory issue and perhaps a goat version of dummy foal syndrome.

As she approached 2 months of age and was still not eating food or drinking water, I called the university vet hospital and talked about bringing her in for an evaluation. It was a hard decision, though, because the prognosis did not sound good. Blindness was the least of her problems, as a goat cannot live on milk forever. If she did not make the transition to eating food and actually being a ruminant, her life expectancy would be very short. Even if they could figure out what was wrong with her, they probably could not do anything to help her survive or become a productive member of our herd. Then a couple of days later, she started to go downhill fast. Over the course of about twenty-four hours, she took less and less from each bottle, and then she died.

6

DEATH

The next post starts out saying that I didn't really want to write it, and even though Coco died seven years ago, it is still a painful memory, so it was hard to include this story. I decided to include it, however, because one of the comments on the original blog post said, "Thank you for sharing this. It helps prepare those of us who have not had to experience it yet, and fortify us for the inevitable." As a wise farmer once said, if you can't stand the thought of an animal dying, don't get any livestock because they all wind up as dead stock someday. Because I was so upset following Coco's death, I really did not give any details in the original blog post about the birth itself, so I'll provide those details now before you read the post I wrote shortly after her death.

Coco was carrying quintuplets, which I didn't know when was in labor. She had given birth to quintuplets before, but because it is so rare and because she was 9 years old, I didn't expect her to be carrying that many kids again. She had been in mild labor all day, and I kept thinking that she'd probably give birth within a couple of hours. But as each hour passed, she did not appear to be in any more pain than the hour before. She kept eating, and she wasn't making any noise. It looked like she was giving a mild push every hour or so. She'd stretch out her neck and curl her lip up while lying down and pushing her hind leg out to the side, but she never seemed to be trying very hard.

By 10:00 in the evening I was worried. I can't explain exactly why, but I felt that something was wrong, so I decided to do an internal check. When I discovered that she was fully dilated, I was briefly relieved, but then I realized that I could feel two kids still in the uterus, ready to be born, but Coco was not even trying to birth them. I could easily identify a nose and two front hooves as belonging to one kid, so I pulled him out. I assumed at that point that the other kids would soon follow. They usually do. Just because one is having trouble coming out doesn't mean the others will. When Coco had quints two years earlier, my daughter Katherine had to pull the first one, but the second one popped out before she had even dried off the first one, and the others quickly followed.

After fifteen or twenty minutes, Coco had not given birth to another kid, and she didn't even seem like she was in labor, so I did another internal check. I discovered two noses and multiple hooves. When I pulled on the two hooves of one kid, the head would fall back over the kid's body. I'd reach in and pull the nose forward, but as soon as I pulled on the feet, the head would fall back again. I didn't have a kid puller that would go around the kid's head to hold it in place while I pulled on the feet, so after trying several times to pull the kid, I knew we had to take her to the university vet hospital.

We arrived at 2:00 in the morning, and the vet on call used a kid puller to get the first kid out. Ultimately, she was able to pull four live, healthy kids. As we all focused on the adorable kids that had just been born, I didn't realize that Coco was not her usual social self. She was staring blankly at the wall. She was ignoring the kids. When I noticed her odd demeanor, the vet said that she was just mad. After the kids had nursed, we loaded them up with Coco and headed home, arriving shortly after 5 a.m.

Because Coco was still acting so strangely, I decided to bring the kids into the house. I was worried that she might fall on them or somehow hurt them, and after everything we had gone through to ensure their safe arrival into the world, that would have been especially upsetting for us. I went to bed and was awakened a couple of hours later by my husband, who told me that he had just found Coco dead. Everyone at the vet hospital was shocked at the news, and they asked me to bring her body back for a necropsy because they couldn't understand why she died. In the end, we learned that she had bled to death from a large tear in her uterus, which probably happened when the kids were being pulled.

Kidding complications that result in death are extremely rare, and even though it was the first time it had happened to us in the eleven years we had been raising goats, it was an event that made me question everything. I second-guessed my decision to take her to the vet, wondering if she would have fared better had I simply kept trying. I felt guilty for not demanding that they simply do a C-section, and for weeks I wondered what I could have done differently, even though the logical part of my brain argued that I had done everything possible. I even questioned my desire to continue raising goats. Ten days after Coco died, I wrote the following blog post with tears streaming down my face.

B·L·O·G
THU
MAR 14
2013

Farewell and thank you, Coco

There are some posts that I don't want to write, and this is one of them, most likely because a part of me would like to deny what actually happened last week and pretend that it didn't happen. But it did happen, and at times like this, I am reminded of Henry David Thoreau's words:

> I wish to live deliberately, to front only the essential facts of life. I wish to learn what life has to teach, and not when I come to die, discover that I have not lived. I do not wish to live what is not life, living is so dear, nor do I wish to practice resignation, unless it is quite necessary. I wish to live deep and suck out all the marrow of life. I want to cut a broad swath, to drive life into a corner, and reduce it to its lowest terms. If it proves to be mean, then to get the whole and genuine meanness of it, and publish its meanness to the world; or if it is sublime, to know it by experience, and to be able to give a true account of it.

On Monday, March 4, 2013, my sweet Coco passed away a few hours after giving birth to quintuplet kids. Mike, Sarah, and I got two hours of sleep the night before because we had taken her to the University of Illinois veterinary clinic when I couldn't untangle the last four kids that were trying to be born. They thought she was fine to leave, but she died less than two hours after we brought her home, and then they asked me to bring her back for a necropsy. They discovered that she had a 14-centimeter tear in her uterus and had bled to death into her abdomen, which was why we didn't see any blood.

It was painful and ugly, or "mean" as Thoreau would say. For the second time in eleven years, I questioned why I'm doing this. I felt horribly guilty. Had we not been breeding our goats so that we could produce our own dairy products—because goats have to give birth to make milk—Coco would not have died. For a brief moment as I was driving down Route 47 to take Coco's body back to the university for the necropsy, a part of my brain said that I couldn't do this any longer. But another part of my brain immediately fired back, *What is the alternative?* Buy dairy products that came from cows injected with hormones, living in factory farms, whose babies are taken away from them at birth? Never! Buy dairy products from small family farms where the animals are treated

After I delivered Coco's first kid at home, she was doing great and helped to clean him off.

humanely? Great idea, but there are limited options here, and in most cases the babies are still taken away. Become a vegan again? Not a bad option, but I love my yogurt and cheese, as well as goat milk in my coffee.

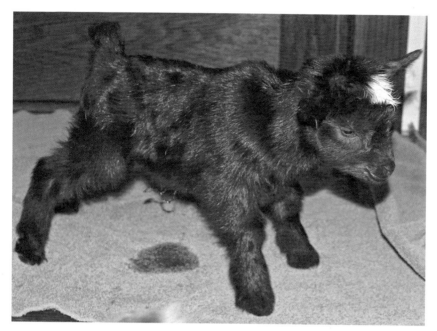

We used to train kids to pee on a towel years ago when we kept them in house longer than they really needed to stay in the house. This is one of Coco's quintuplets. Teaching them to pee on a towel is simple. Just hold the kid for about fifteen or twenty minutes after giving them a bottle, then stand them on a towel. They'll usually pee right away. If they start to pee somewhere else, just move them to the towel quickly. The only downside to this is that they don't know the difference between a towel and other fabrics or even throw rugs, so they need to be in a room with a hard floor and nothing else soft on the floor other than the towel you want them to use. You cannot, however, teach them to poop anywhere in particular. Luckily, newborns poop very little, and it doesn't stink. It looks like scrambled eggs and is very easy to pick up with a piece of toilet tissue. Like human babies, their poop doesn't turn brown and stinky until they start eating solid food.

A lot of people say that when their goat has multiples, one or more of the kids are quite small, but all of Coco's quints were normal size, averaging about 3 pounds each.

In our modern world, we are protected from so many of the essential facts of life—far more than Thoreau could have ever dreamed. People in our modern world are so oblivious to the simple facts of life because the reality is all hidden away in factories and hospitals and other institutions. I couldn't tell you how many people I've met over the years who have no clue that cows are bred every year to continue producing milk in factory farms, or how many people think that white chickens lay white eggs and brown chickens lay brown eggs. They apparently have no idea that there are gold-and-black-laced chickens and black-and-white-barred chickens and so on! The average life of a factory-farm cow is only about four years, even though cows can live to be fifteen or more. And factory-farm hens, which are debeaked, are turned into odd chicken bits after little more than a year of life, which is spent inside a wire cage, so they never see the sky, run across grass, or catch even a single bug during the unnaturally short life.

When people have no clue about what is normal or possible, it makes it very easy for advertisers to convince them that things like confinement buildings, daily antibiotics, and debeaking are for the animal's own good. When we moved out here to grow our own food organically in 2002, my knowledge about our modern food system was a tiny fraction of what it is today. And with what I know today, I am more committed than ever to continuing this life, even knowing that sometimes it will get mean and ugly. Life is not a perfect, shiny, cellophane-wrapped package. Nor is it a dinosaur-shaped "chicken" nugget. Milk and meat and vegetables do not come from a store. Those are modern illusions.

Life is a chicken running through the grass, catching bugs, laying eggs, and sitting on those eggs until they hatch, bringing forth more chickens that will grow up and lay eggs or become a chicken dinner. Life is a sheep grazing in the pasture for a year to bring forth a few pounds of wool. Life is a turkey running from a coyote and flying up into a tree so that we can have a turkey dinner. Life is a garden that is filled with bugs, both good and bad, that will help and hinder us in every step that gets us closer to harvest. Life is a goat waddling around when she's pregnant and screaming through labor contractions to bring forth kids that will tell her body it's time to make milk to feed them. And life always ends in death.

In her nine years of life, Coco Chanel gave us twenty-seven kids and hundreds of gallons of milk. I think of her every day when I see her daughters Vera Wang and Nina Ricci. And I'm sure I'll think of her often as her newborn Bella Freud grows up and becomes a mother and a milk goat. I can point to aging blocks of cheddar and gouda that include milk that she produced, which we'll be eating in the years to come. Coco was an amazing mother, growing big babies, even when there were four or five, giving birth to them, and then nursing them. Even though she was carrying five babies this time, they were all the same size as normal twins would have been.

And unlike cows in factory farms that produce the majority of dairy products in this country, Coco was loved and appreciated. She had a name, not an ID number, and she had a personality that set her apart from the other goats on the farm.

I'll always remember Coco as my baby, the one who refused to grow up. She tried to die on me when she was only two weeks old, but I wouldn't let her. I remember holding her in my arms on the couch, crying, "Please don't die." It was only our second year out here, and I hadn't seen an animal die yet. In spite of my inexperience, we pulled her through. I called her PeeWee, and she wound up as a bottle baby. Katherine took her to 4-H meetings, where everyone cooed over the tiny little brown doe.

When we realized she really was going to live, we named her Coco Chanel, partly because she was chocolate colored and partly because she needed a classy name to suit her. Regardless of how old she got, though, she always thought she was a lap goat. Whenever I sat down anywhere near her, she would walk over and try to crawl into my lap. Even as we were driving to the vet hospital last week, she was trying to edge her over-sized, very pregnant body into my lap as she and I were in the back of the car together.

As hard as this has been, I wouldn't trade the last nine years with Coco for anything, and although this was the first time we had a goat die as a result of kidding, I know it won't be the last… because death is an inescapable part of life.

Our most challenging birth… with a happy ending

WED FEB 17 2016

Monday afternoon Sadie finally went into labor. This was after a week of our being on kid watch and thinking that she or Cicada were ready to give birth. I was supposed to be teaching a class at the local community college on how to get your writing published. My husband, Mike, and intern Stefanie were here, and although part of me thought that it would be fine to leave them in charge, there was another part of me that thought something wasn't quite right.

Sadie started pushing shortly after 4:00. She wasn't pushing particularly hard and didn't seem distressed, but she kept changing positions—sitting like a dog, standing and arching her back up like a camel, or squatting to push. Everything in my head said that the kid was not in a great position and she was changing positions to try to move it along. Most goats give birth lying on their side, and those that do give birth standing are not making the back look like the classic Halloween cat.

At 4:30 I was thinking that I should be going into the house to shower and have dinner, but surely the first kid will be born any minute now, and I can show Stef how to dry it off so that she can handle the rest as they're born. This was her first birth.

By 5:00 I was thinking that I didn't really need that much time to shower and eat and that that first kid was going to be here really soon, especially after I saw a bubble of water appear and break. I slipped one finger into Sadie and felt something hard, pointy, and boney. *It's probably a butt. It's definitely not a nose.* Breech kids take a little longer to push out because the butt is blunter than the nose, which is torpedo-shaped, but this kid was sure to be born soon.

As the minutes ticked past, I started to think that I can get away without showering. More time passed. I told Stef that I didn't need dinner, that I could just eat a banana in the car as I was driving to class, and that I'd eat dinner when I got home at 9:30. When I checked Sadie again, I immediately realized the kid had not budged at all in the last half hour. Something was definitely not right. I realized I'd have to stay with Sadie. Mike had never dealt with any labor complications, and his hands are much bigger than mine.

I called the college and apologized profusely for canceling at the last minute. "I have a goat in labor who is having complications, and I can't leave."

Sadie was pushing so hard that her rectum was prolapsing. Basically she was pushing it inside out. It's not something I'd seen before, and I hope to never see it again. Thankfully, when she stopped pushing, everything went back inside where it belonged. It happened a few more times, but it all went back inside whenever she stopped pushing.

I ultimately figured out that the kid was indeed breech, but unlike every other breech kid born on this farm, whose legs were straight up against its stomach and chest, this little darling had its legs folded as if it were lying in the pasture enjoying a sunny day. So, in your typical breech, there is only the circumference of the kid's butt and a single set of bones from its hind legs that are pressed straight against its body. But in this case, there was the circumference of the kid's butt PLUS three sets of bones from the hind legs that were folded up like a Z against its body.

Summarizing the whole thing in a paragraph like that makes it sound far simpler than it was. I spent a lot of time trying to figure out exactly what I was feeling when I examined Sadie, and I was talking out loud to Mike and Stef about it, trying to help myself visualize it all. I knew I needed to find the hind feet and pull them out, but I couldn't find them. Finally, I decided to phone a friend who has a lot more goats than I do and so has seen a lot more unusual presentations. As I was walking towards the house, my brain seemed to connect the dots of what I had been feeling inside Sadie, so I was able to explain it to my friend far better than I could have a few minutes earlier.

She said I needed to push the kid back inside so that I could straighten out the back legs and pull them out of the doe to deliver the kid. "Do you have a cattle sleeve," she asked, referring to the plastic gloves that go up to your shoulder. "Yes," I replied.

"If you're wearing a shirt, take it off because it's just going to get in your way—unless you can push your shirtsleeve up to your shoulder." I didn't bother telling her that it was 30°F here or that I was actually wearing a shirt, a sweater, and a coat. None of that changed what needed to be done.

When I got back to the barn, I explained to Mike and Stef what I needed to do. I took off my coat, trying to ignore the cold, and managed to push my shirt and sweater up above my elbow. Stef asked me a question about the kid's presentation and how I would be rearranging it, and as I responded to her question, I actually clarified for myself what I was doing. Although the concept was simple, the task itself was not easy, and it was punctuated with me saying, "I hate this, I hate this," and "I'm so sorry, Sadie," over and over again. Once I had both legs out, it really was not hard at all to pull the kid out. It was a doe! And she didn't seem to be the least bit troubled by the ordeal that her mother and I had endured to get her into this world.

As we were cleaning her up, Sadie pushed out the second kid, which was another doe, also in excellent condition.

Then a big bubble of fluid appeared under Sadie's tail. I saw something small and black in the bubble, and I said, "That looks like an ear." Then in a moment of wishful thinking, I said, "It's probably a tail," even though my brain was saying that a tail would be shorter and thicker. It was indeed an ear, which, unfortunately for Sadie, meant that I would have to push the kid back inside her, where there would be more room, so that I could flip up the chin so that it could come out nose first. Having just rearranged the first kid, I was not feeling nearly as nervous about helping this one.

We ultimately learned that the third kid was the largest of the three—also a doeling—at 4 pounds, 2 ounces. The first one weighed 3 pounds, 8 ounces, and the second one was 3 pounds, 6 ounces. By the time we got all of the kids to nurse, it was 9:00, which was when my class at the community college would have ended. I was glad that I had canceled, because I probably would have been called home early if I had gone in. Or, if no one had called me, we probably would not have had three live, healthy doelings.

One of the best lessons from this story is that if someone had asked me what to do in this situation, I could have told them. However, when I was in this situation, I couldn't figure out what to do because I wasn't thinking straight. As soon as my friend told me what to do, I thought, *Duh! I knew that!* But it also helped to talk through all of it. This is why I recommend that you always have a phone number handy for a vet as well as for a knowledgeable goat mentor who can help you figure out what to do.

It's incredibly ironic that I titled this post "with a happy ending" because the story was not over. Five weeks later, I wrote another post about Sadie.

Some decisions just suck

As much as I love my homesteading lifestyle, there are times when it just sucks. Now is one of those times.

Last week, our intern told me that Sadie was crying out when she was peeing. I went to the barn and watched her, and within a couple of minutes, she was squatting and screaming, and only a few drops of pee came out. I was puzzled, as I have never heard of a doe having a urinary stone.

I called the University of Illinois veterinary clinic, and of course they suggested I bring her in, which I did. They ran a lot of urine and blood tests, which all came back normal, which meant she did not have an infection. She also did not have a fever. The interesting thing was that whenever anyone put pressure on her bladder—whether under her belly or on her sides—pee would squirt out of her like a water hose on high! They did an ultrasound, which showed an abnormal uterus, and Sadie had given birth five weeks earlier. They hypothesized about the contents of the uterus, and I learned a lot listening to their brainstorming. But they said we could learn more about the contents of the uterus by doing a CT scan. It would be about $450, and our bill was already up to a couple of hundred. Even though the solution might lie in a hysterectomy, which would mean she could never have kids again, I agreed to do the test.

It showed that Sadie's uterus was filled with fluid. They were fairly sure that it wasn't pus because of the color, that it was probably blood or some other type of fluid. There are only two ways to get the fluid out. One is to give her a shot of prostaglandin, which should hopefully cause her to go into heat, which would open her cervix so it could drain. The other way to drain her uterus is surgically—make a small incision and suction it out. However, will it fill up again?

I opted for the prostaglandin, and they gave her an injection on Saturday. As of Monday, she had not come into heat, so they gave her another injection at a higher dose. As of Tuesday afternoon, she had not come into heat, so they're planning to give her another injection today, Wednesday. But they said we should probably start thinking about surgical options. Other than draining her uterus, which may not work long term, a complete hysterectomy is the best way to eliminate the problem.

But then I'll have a dairy goat who can never get pregnant again. So, we could milk her until she dried up from this lactation, and then she would be worthless as a dairy goat. The bill is already up to $1,100. Surgery would be another few hundred.

This is when you realize that being a grown-up is not as awesome as you thought it was going to be when you were 12. This is when you wish that you had enough money that a couple of thousand dollars didn't mean anything to you. This is when life sucks.

I want Sadie to be able to pee on her own and to come home and to continue nursing her three beautiful doelings that have already grown so big and strong during their first five weeks on their mama's milk. Since Sadie got sick, we've been trying to feed them milk from the other does with a bottle, and some feedings go better than others. Sometimes they seem to get it, and sometimes they have no clue. They're getting only about 50 percent as much as they should every day. Even though they're eating hay and grain, their immune systems are still very immature, and with the stress of losing their mama, I'm very concerned that I'll soon be seeing poopy butts caused by coccidiosis.

What if we pay for the surgery, but when Sadie comes home, she no longer remembers her kids and won't let them nurse? What if the stress of all this causes her milk to dry up? Or, what if she dies in spite of the surgery—or because of it?

I suck as a business person. A good business person would have said to euthanize last week as soon as this became complicated. A good business person doesn't lose sight of the bottom line. But I am clearly not a good business person. The bill is sitting at $1,100 now, and I still can't give up on her, at least not today. She is still at the vet hospital, and I'm still hopeful that another shot of prostaglandin will do the trick. But what will I do if the prostaglandin doesn't work? I know what a good business person would do. I know what my heart wants me to do. But I honestly don't know what I will do.

WED
MAR 30
2016
(later in
the day)

Farewell, Sadie

The vet called today shortly before noon to say that she had consulted with a couple of reproductive specialists and they said that if Sadie has not responded to the first two shots of prostaglandin, there is really no point in making her miserable with a third. If her cervix is not dilating, it's because of scar tissue or adhesions, and it's not going to dilate.

They did not recommend draining the uterus surgically because if it was full of pus, they would not be able to keep it from the abdominal cavity, so it would cause more problems. They did not recommend draining it through the cervix with a trochar because it would leave her with less than a 5 percent chance of ever carrying a pregnancy to term, and the odds are high that the uterus would simply fill with fluid again.

They said that if her uterus continued to fill with fluid, it would eventually cause her to be unable to poop, and she was already straining to have a bowel movement. The uterus could eventually rupture, if she lived long enough. She's had a catheter in her for as long as she's been there so that her bladder won't overfill again, but they couldn't send her home with a catheter.

In the end, our options came down to euthanasia or a hysterectomy, which would cost $1,000, assuming nothing went wrong, which could drive up the price even more. Apparently doing a hysterectomy on a goat is more complicated than on dogs and cats. When I left her there on Thursday, the estimate was $700 to $800, and the bill is already $1,100.

It's been a horrible few hours, thinking about what to do. Hearing the estimate for a hysterectomy came as quite a shock. Last week, they had said it probably would be a few hundred dollars. A thousand dollars is about three times as much as what I'd been preparing myself for. I just can't justify spending $1,000 to keep her alive, knowing that she would "only" be a pet afterwards. I already have four retired does, as well as an 8-year-old that has never been bred because she never grew big enough. Still, I hate the idea of essentially putting a price tag on her life. If I had all the money in the world, I would have easily said to do the hysterectomy. But I can't live my life as if I have money I don't have. And what would I do if next week another animal were to get injured or become ill? I have to try to maintain some sort of fiscal responsibility. Still, I hate it.

It makes it even worse for me that she's spent the last six days of her life alone in a strange place with strangers who've been poking her with needles and a catheter. I justified leaving her there because I thought they could make her well and she'd be able to come home and be with her babies again. Sadie was such a shy kid, but after she freshened, she warmed up to me and the rest of the humans. I always told her that she could trust us, and as I was leaving her at the vet hospital last week, I couldn't help but think that she felt betrayed. If only I could have explained to her what was happening…

So, I called the vet a little while ago and told her to euthanize Sadie. It was one of those horrible moments in life. I hated it. But

I also hated that she had endured six more days of suffering because we were hoping to save her. They had said the prostaglandin injections might not work, but in my head, I was thinking that they had to work. She would go into heat; her cervix would open; and her uterus would drain. It sounded so simple. Surely, Sadie would be coming home again.

But then I think about why farmers tend to be such spiritual people. Maybe it's to protect our sanity, but we have to believe that everything works out for the best. Perhaps Sadie was with us for only four years because that's what was meant to be. She's given me all that I was meant to have from her. I just realized that I have her son from last year. It sounds silly, but I kept forgetting to advertise him. Maybe there was a reason for that. This year, the person who had wanted two of Sadie's doelings wound up having to cancel after I had already marked them as sold on the website, and Sadie got sick before I had a chance to update the website to state that they were available.

Still, it's always hard to lose a good goat, and Sadie was great. She was very generous about giving me plenty of doe kids, starting with triplet does when she was only a year old. In four years, she gave us fourteen kids, which included a remarkable eleven does! She also had outstanding parasite resistance, never having had a dewormer in her life. And until this year, she never had any birthing problems, even shooting out a couple of breech kids without help as a first freshening yearling. I'm so grateful to have three of her kids here, especially since I never had the foresight to consciously keep any of them, thinking that I had plenty of time since she was only 4 years old. Although we won't see her sweet face again, she left behind a great legacy in her kids. Perhaps that's the best thing that any of us—human or goat—can ever hope for, regardless of how many years we live.

FINAL THOUGHTS

By now you may have realized that I never say "never" when talking about what can happen in a birth. However, there is one time that I say "never," and that is when it comes to how you react. Never panic! That is much easier said than done, and it took me at least a decade to remember that when I'm in the midst of a birth that becomes challenging. Sometimes we think that something must be done right this second to save a kid, but that is actually not the case if the kid is still entirely inside of the doe. Kids do not go from perfectly healthy to dead within a few seconds.

Unfortunately, we don't know when the kids are starting to get stressed. It seemed that Giselle was not in labor nearly as long as Caboose or Coco, yet Giselle's kids were born dead. I am two hours from the university vet hospital, and in eighteen years I've had to take only four goats there while they were in labor, and only four kids were born dead, compared to eleven that were born alive. My friend who had to drive two hours to take her Nubian down there for a C-section also wound up with four live kids.

Although I mentioned using Yahoo groups in my early years, I do not recommend getting advice from social media groups today, especially in an emergency situation. Back then there were only a few hundred people in a group, and everyone knew who was a novice and who was experienced enough to be offering advice. Today's groups have tens of thousands of members who are mostly total strangers. Even in my small groups years ago, there were times when people misunderstood or didn't ask enough questions or simply panicked when they read a post. I did not always get good information, and sometimes I made mistakes as a result. As groups have grown in size, the newbies have started to outnumber the experienced people, and as more people respond with a variety of opinions, it gets more confusing for the person asking the question.

I know it's not always easy to find a goat vet in many parts of the country, but it is absolutely vital that you find one before you bring home your first goats. There will be times when you need their medical expertise, diagnostics, equipment, and prescription drugs. When talking to one vet about the most common mistakes people make with their goats in labor, she gave two seemingly opposite answers. First, she said that most

people intervene too soon. And second, she said that most people don't call the vet soon enough for them to be able to help in a true emergency situation. If you think something is wrong, call your vet and talk it over. They will either give you reassurance that your goat is fine, or you will be moving closer to actually getting your goat the help it needs.

So, remember, you have nothing to gain by panicking. And you could make a big mistake, as I did when Giselle had a dead kid that was presenting ribs first. If that kid had been alive, I could have done some serious damage to its leg. Now, when I feel like I am about to panic, I take a deep breath and remind myself that seconds will not matter. I have time to assess the situation and think about what to do. And sometimes that

means calling someone else for advice, whether it is a knowledgeable friend or a veterinarian.

But above all else, remember that goat births will almost always be perfectly normal. The hardest thing you'll have to do is dry off the kids. Then you can coo over how adorable they are, take lots of pictures, and soak in all of the compliments on social media as people gush over their cuteness.

Index

In this index, page numbers set in **Bold** indicate an illustration.

About the Author

DEBORAH NIEMANN and her family moved to the country in 2002 to start producing their own food organically, including their own goat cheese. Before she knew what happened, two milk goats turned into 20, and a desire to make a simple chèvre launched a whole new career helping people raise their goats. Deborah is the author of *Homegrown and Handmade*, *Ecothrifty*, and *Raising Goats Naturally*. She teaches online courses on raising goats and chickens at the University of Massachusetts and contributes to magazines such as *Hobby Farms*, *GRIT*, *Mother Earth News*, *Chickens*, and *Urban Farm*. Deborah maintains a homesteading blog from her farm in Cornell, Illinois: thriftyhomesteader.com.

ABOUT NEW SOCIETY PUBLISHERS

New Society Publishers is an activist, solutions-oriented publisher focused on publishing books for a world of change. Our books offer tips, tools, and insights from leading experts in sustainable building, homesteading, climate change, environment, conscientious commerce, renewable energy, and more—positive solutions for troubled times.

We're proud to hold to the highest environmental and social standards of any publisher in North America. When you buy New Society books, you are part of the solution!

- We print all our books in North America, never overseas

- All our books are printed on **100% post-consumer recycled paper**, processed chlorine-free, with low-VOC vegetable-based inks (since 2002)

- Our corporate structure is an innovative employee shareholder agreement, so we're one-third employee-owned (since 2015)

- We're carbon-neutral (since 2006)

- We're certified as a B Corporation (since 2016)

At New Society Publishers, we care deeply about *what* we publish—but also about *how* we do business.

Download our catalog at https://newsociety.com/Our-Catalog or for a printed copy please email info@newsocietypub.com or call 1-800-567-6772 ext 111.

ENVIRONMENTAL BENEFITS STATEMENT

New Society Publishers saved the following resources by printing the pages of this book on chlorine free paper made with 100% post-consumer waste.

TREES	WATER	ENERGY	SOLID WASTE	GREENHOUSE GASES
24	2,000	10	80	10,370
FULLY GROWN	GALLONS	MILLION BTUs	POUNDS	POUNDS

Environmental impact estimates were made using the Environmental Paper Network Paper Calculator 4.0. For more information visit www.papercalculator.org.

Certified

Corporation

MIX
Paper from responsible sources
FSC® C016245

new society
PUBLISHERS
www.newsociety.com